App 自动化测试与框架实战

刘金起　李明黎◎著

人民邮电出版社

北京

图书在版编目（ＣＩＰ）数据

App自动化测试与框架实战 / 刘金起，李明黎著. --北京：人民邮电出版社，2019.1
　ISBN 978-7-115-49121-3

　Ⅰ．①A… Ⅱ．①刘… ②李… Ⅲ．①移动终端－应用程序－程序测试 Ⅳ．①TN929.53

　中国版本图书馆CIP数据核字(2018)第185407号

内 容 提 要

本书从 App 测试基础、实战技术，再到自动化测试框架的搭建，全面地讲解 App 测试所需要的知识，主要内容为：App 测试及其类型、Java 编程环境构建、Java 语言基础、Android 自动化环境精讲、Android 自动化测试基础精讲、Android Appium 自动化框架、Appium 数据驱动测试框架封装实战、Appium 关键字驱动测试框架封装实战、持续集成的自动化、Appium 常见问题处理方式。

本书适合测试初学人员、测试工程师、质量管理人员阅读，也适合作为大专院校相关专业师生的学习用书和培训学校的教学用书。

◆ 著　　刘金起　李明黎

责任编辑　张　涛

责任印制　焦志炜

◆ 人民邮电出版社出版发行　北京市丰台区成寿寺路11号

邮编　100164　电子邮件　315@ptpress.com.cn

网址　http://www.ptpress.com.cn

固安县铭成印刷有限公司印刷

◆ 开本：800×1000　1/16

印张：24.5　　　　　　　　　2019年1月第1版

字数：484千字　　　　　　　2024年7月河北第5次印刷

定价：79.00 元

读者服务热线：(010)81055410　印装质量热线：(010)81055316
反盗版热线：(010)81055315
广告经营许可证：京东市监广登字20170147号

前言

杜绝简单、呆板的纸上谈兵，强化实战经验分享，是本书的写作初衷。

全书分为 11 章，用层层递进的方式帮助读者完成 App 测试分析、进行 App 自动化测试实战，并搭建 App 自动化测试框架，全方位帮助读者学习 App 测试知识。书中也包含了作者十几年来在持续集成领域的心得体会，并专门安排一章讲述持续集成自动化测试。本书知识架构如图 0-1 所示。

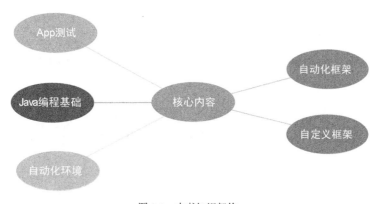

图 0-1 本书知识架构

本书全方位呈现了 App 测试的方法和技术。第 1 章和第 2 章主要讲解 App 产品是如何测试的。不管是自动化测试还是手工测试，测试的基本着手点是不变的，也就是说，我们的测试策略、测试方案的思考维度是统一的，这两章结合起来就是一个 App 测试策略分析点的 Check List，自动化测试和手工测试只是用来实现测试所执行的两种不同手段，而这些测试的分析过程是一致的。

这两章也清晰表述了 App 中哪些测试必须借助自动化手段，哪些测试使用自动化测试手段会得到更好的结果，并实现更高的效率，方便从事自动化测试工作的读者在做自动化测试可行性分析时参考。

第 3 章和第 4 章通过简单明了的案例，教会读者快速搭建 Java 测试环境并掌握 Java 语言的编程。我们抽取了自动化测试中常用的概念、编程封装技巧，通过重点概念理论分析和精心设计的例子，让读者用较少的时间、较小的精力，掌握常规自动化测试的使用知识。

第 5 章和第 6 章详细讲解了 Android 自动化测试的基础环境。众所周知，Java 是实现 Android 自动化测试的编程语言，它的编程环境一向比较复杂。这里仔细讲解相关准备工作，让读者清晰地把握其脉络，并轻松搭建起自动化测试环境。这里特意安排了 Android 系统架构的讲解，要深入进行 Android 自动化测试，就必须了解其机制，了解其系统架构。这也是我们自己编写测试框架的必备知识。

第 7 章讲解了本书的核心知识——Appium 框架，这也是本书的一个难点。这一章的讲解力求简约而清晰，使读者学以致用。有了上面的这些知识储备，接下来我们就可以准备封装自己的自动化测试框架了。

第 8～11 章是本书的实战部分。第 8 章讲解了主流的数据驱动自动化测试框架，第 9 章讲解比较高端的关键字驱动框架。第 10 章讲解了如何使自己编写的测试框架实现持续集成自动化，这是自动化测试中一个里程碑式的工作，在公司里能够实现自动化的持续集成几乎是每个自动化测试团队的一个终极目标。第 11 章分享了作者的自动化测试经验，对于常见自动化测试框架搭建中的问题进行汇总，并给出了实战中一些问题的解决方案。

自动化测试框架的实现思想是相通的，只是需要反复实践、认真思索，希望读者在本书的基础上迅速学会自动化测试框架的搭建技巧，并设计和开发出更优秀、更高效的自动化测试框架。

本书不仅可供希望学习自动化测试的读者参考，刚刚开始尝试做 App 手工测试的读者同样可以借鉴本书。

本书力求简单明了，分享了大量经验和技术"干货"。尽管本书准备已有时日，且经反复雕琢，最后两个月更是通过作者、热心朋友逐字逐句地校对修改，每一个架构图都精心推敲，反复讨论修正，甚至章节和段落间的过渡文字都反复打磨，但终究作者水平有限，书中仍难免有些不足之处，恳请读者批评指正。

在这里真心地感谢人民邮电出版社的编辑张涛，这段时间里一直支持并鼓励着我们的写作，正因他对本书的指导，才能让本书快速地与读者见面。

最后，谨以本书献给测试界的广大同仁和即将踏入测试之门的朋友，希望大家都能在测试岗位上越做越好！

本书编辑和投稿联系邮箱为 zhangtao@ptpress.com.cn。

刘金起

资源与支持

本书由异步社区出品，社区（https://www.epubit.com/）为您提供相关资源和后续服务。

配套资源

本书提供配套的源代码。

要获得以上配套资源，请在异步社区本书页面中单击 配套资源 ，跳转到下载界面，按提示进行操作即可。注意，为保证购书读者的权益，该操作会给出相关提示，要求输入提取码进行验证。

如果您是教师，希望获得教学配套资源，请在社区本书页面中直接联系本书的责任编辑。

提交勘误

作者和编辑尽最大努力来确保书中内容的准确性，但难免会存在疏漏。欢迎您将发现的问题反馈给我们，帮助我们提升图书的质量。

当您发现错误时，请登录异步社区，按书名搜索，进入本书页面，单击"提交勘误"，输入勘误信息，单击"提交"按钮即可。本书的作者和编辑会对您提交的勘误进行审核，确认并接受后，您将获赠异步社区的 100 积分。积分可用于在异步社区兑换优惠券、样书或奖品。

扫码关注本书

扫描下方二维码，您将会在异步社区微信服务号中看到本书信息及相关的服务提示。

与我们联系

我们的联系邮箱是 contact@epubit.com.cn。

如果您对本书有任何疑问或建议，请您发邮件给我们，并请在邮件标题中注明本书书名，以便我们更高效地做出反馈。

如果您有兴趣出版图书、录制教学视频，或者参与图书翻译、技术审校等工作，可以发邮件给我们；有意出版图书的作者也可以到异步社区在线提交投稿（直接访问 www.epubit.com/selfpublish/submission 即可）。

如果您是学校、培训机构或企业，想批量购买本书或异步社区出版的其他图书，也可以发邮件给我们。

如果您在网上发现有针对异步社区出品图书的各种形式的盗版行为，包括对图书全部或部分内容的非授权传播，请您将怀疑有侵权行为的链接发邮件给我们。您的这一举动是对作者权益的保护，也是我们持续为您提供有价值的内容的动力之源。

关于异步社区和异步图书

"异步社区"是人民邮电出版社旗下 IT 专业图书社区，致力于出版精品 IT 技术图书和相关学习产品，为作译者提供优质出版服务。异步社区创办于 2015 年 8 月，提供大量精品 IT 技术图书和电子书，以及高品质技术文章和视频课程。更多详情请访问异步社区官网 https://www.epubit.com。

"异步图书"是由异步社区编辑团队策划出版的精品 IT 专业图书的品牌，依托于人民邮电出版社近 30 年的计算机图书出版积累和专业编辑团队，相关图书在封面上印有异步图书的 LOGO。异步图书的出版领域包括软件开发、大数据、AI、测试、前端、网络技术等。

异步社区

微信服务号

目录

第1章 了解 App 测试 ·········· 1
  1.1 App 测试与普通软件测试的差异 ·········· 2
  1.2 App 测试的难点 ·········· 4
  1.3 App 测试中的网络信号概述 ·········· 4
  1.4 智能终端中的 App 测试 ·········· 5

第2章 App 测试类型 ·········· 7
  2.1 功能测试 ·········· 8
    2.1.1 高级别事件响应 ·········· 8
    2.1.2 第三方应用打断 ·········· 8
    2.1.3 通信录的备份恢复功能 ·········· 9
    2.1.4 手机和其他外设产品的互联互通 ·········· 9
  2.2 稳定性测试 ·········· 9
  2.3 兼容性测试 ·········· 11
    2.3.1 手机品牌 ·········· 11
    2.3.2 硬件种类 ·········· 11
    2.3.3 芯片种类 ·········· 12
    2.3.4 分辨率 ·········· 13
    2.3.5 各种无线网络的兼容性 ·········· 13
    2.3.6 第三方软件兼容性 ·········· 13
  2.4 性能测试 ·········· 14
  2.5 网络测试 ·········· 14
    2.5.1 室内网络测试 ·········· 14
    2.5.2 外网测试 ·········· 14
    2.5.3 弱场测试 ·········· 15
  2.6 异常测试 ·········· 15
  2.7 发布测试 ·········· 16
  2.8 用户界面测试 ·········· 16
    2.8.1 图形测试 ·········· 16
    2.8.2 内容测试 ·········· 17
  2.9 冲突测试 ·········· 17
    2.9.1 按键打断 ·········· 17
    2.9.2 程序后台相互切换 ·········· 18
    2.9.3 网络切换 ·········· 18
    2.9.4 待机唤醒 ·········· 18
  2.10 接口测试 ·········· 18

第3章 Java 编程环境构建 ·········· 20
  3.1 安装 JDK 与配置环境变量 ·········· 21
    3.1.1 下载 JDK ·········· 21

3.1.2　安装 JDK ········· 22
　　3.1.3　配置环境变量 ········· 24
3.2　安装与配置 Eclipse ········· 28
　　3.2.1　安装 Eclipse ········· 28
　　3.2.2　Eclipse 常用配置 ········· 28

第 4 章　Java 语言基础 ········· 30
4.1　Java 简介 ········· 31
4.2　第一个 Java 应用项目 ········· 31
4.3　函数 ········· 32
4.4　类 ········· 36
4.5　包 ········· 38
4.6　语句 ········· 42
　　4.6.1　条件判断 ········· 43
　　4.6.2　循环判断 ········· 45
4.7　Java 调试技巧 ········· 46

第 5 章　Android 自动化环境精讲 ········· 48
5.1　安装 Android SDK ········· 49
5.2　Maven 项目管理 ········· 54
　　5.2.1　安装 Maven ········· 54
　　5.2.2　安装 Maven 插件 ········· 56
　　5.2.3　创建 Maven 项目 ········· 57
　　5.2.4　Maven 项目依赖包 ········· 60
　　5.2.5　Maven 坐标定位 ········· 61
5.3　TestNG 测试框架简介 ········· 62
　　5.3.1　安装 TestNG ········· 62
　　5.3.2　TestNG 测试用例 ········· 65
　　5.3.3　数据驱动 ········· 72
　　5.3.4　分组测试 ········· 82
　　5.3.5　按照特定顺序执行测试用例 ········· 86
　　5.3.6　忽略测试 ········· 88
　　5.3.7　依赖测试 ········· 89
　　5.3.8　超时测试 ········· 90
　　5.3.9　异常测试 ········· 91
　　5.3.10　测试报告 ········· 93
　　5.3.11　断言 ········· 94
　　5.3.12　通过 Maven 执行 TestNG 测试用例 ········· 95
5.4　Log4j 日志 ········· 97
　　5.4.1　Log4j 安装 ········· 98
　　5.4.2　Log4j 配置文件 ········· 100
　　5.4.3　Log4j 引用 ········· 103

第 6 章　Android 自动化测试基础精讲 ········· 105
6.1　adb 命令 ········· 106
　　6.1.1　在手机上启动 USB 调试 ········· 106
　　6.1.2　adb 命令环境搭建 ········· 107
　　6.1.3　adb 组织结构简介 ········· 108
　　6.1.4　adb 常用命令 ········· 109
　　6.1.5　adb 端口冲突问题解决 ········· 122
6.2　Android 简介 ········· 123
　　6.2.1　Android 常规动作 ········· 124
　　6.2.2　Android 的按键和 Keycode ········· 126
　　6.2.3　Android 坐标点简介 ········· 130
6.3　Android 自动化测试前的准备 ········· 131
　　6.3.1　布局 ········· 131
　　6.3.2　Android 的组件 ········· 131
　　6.3.3　组件属性 ········· 132
　　6.3.4　确定包名和 Activity 值 ········· 132

第 7 章　Android Appium 自动化框架 138

- 7.1　Appium GUI 简介 139
- 7.2　Appium 架构详解 139
- 7.3　Appium Windows 环境搭建 141
  - 7.3.1　Node.js 的安装 142
  - 7.3.2　.NET Framework 的安装 146
  - 7.3.3　Appium 的安装与配置 148
- 7.4　Appium GUI 详解 151
- 7.5　新会话窗口 154
- 7.6　在 Appium 中查找控件 157
  - 7.6.1　Appium Inspector 界面 158
  - 7.6.2　Selected Element 面板 ... 159
  - 7.6.3　操作区域 161
  - 7.6.4　调试定位方式 162
- 7.7　Appium 录制功能 165
- 7.8　Desired Capabilities 的配置 168
  - 7.8.1　Desired Capabilities 配置简介 168
  - 7.8.2　Desired Capabilities 配置示例 171
- 7.9　识别对象的 API 方法 175
  - 7.9.1　通过 Name 属性识别 175
  - 7.9.2　通过 ClassName 属性识别 176
  - 7.9.3　通过 Id 属性识别 177
  - 7.9.4　通过 AccessibilityId 识别 178
  - 7.9.5　通过 XPath 识别 179
  - 7.9.6　通过 UIAutomator 识别 ... 183
  - 7.9.7　通过 cssSelector 识别 184
  - 7.9.8　通过 LinkText 识别 187
  - 7.9.9　通过 PartialLinkText 识别 187
  - 7.9.10　通过 TagName 识别 188
  - 7.9.11　通过 by 类识别 188
  - 7.9.12　通过 getPageSource 识别 189
  - 7.9.13　通过坐标界定对象识别 189
  - 7.9.14　按照权重识别 191
- 7.10　其他 API 方法详解 192
  - 7.10.1　与控件信息相关的 API 方法 192
  - 7.10.2　与手势相关的 API 方法 193
  - 7.10.3　与 TouchAction 相关的 API 方法 193
  - 7.10.4　与系统操作相关的 API 方法 195
- 7.11　Android 测试实例 196
  - 7.11.1　Android 原生 App 实例 196
  - 7.11.2　Android 移动 Web App 实例 200
  - 7.11.3　Android 混合 App 实例 205
- 7.12　查看 Appium 日志 212

第 8 章　Appium 数据驱动测试框架封装实战 236

8.1	自动化测试规划与设计	237
8.2	配置Maven与创建Maven项目	238
	8.2.1 配置Maven	239
	8.2.2 创建Maven项目	243
	8.2.3 Maven项目依赖包	243
8.3	配置Git	244
8.4	配置SVN	245
8.5	TestNG工具	245
8.6	配置Appium	246
	8.6.1 在Maven中导入Appium	246
	8.6.2 创建测试脚本	247
8.7	设计模式	252
	8.7.1 PO模式	252
	8.7.2 PageFactory模式	257
8.8	数据驱动	263
8.9	公共库	271
8.10	Log4j日志	276
	8.10.1 在Maven中导入Log4j	276
	8.10.2 Log4j的使用	277
8.11	ReportNG测试报告	289
	8.11.1 通过Maven导入ReportNG	289
	8.11.2 配置ReportNG的监听器	290
	8.11.3 执行测试	291
8.12	Appium自启动	293

第9章 Appium关键字驱动测试框架封装实战 296

9.1	搭建测试框架	297
9.2	代码优化	298
9.3	关键字驱动	300
9.4	页面元素的封装	307
9.5	测试操作的封装	310
9.6	执行测试	318

第10章 持续集成的自动化 325

10.1	安装Jenkins	326
	10.1.1 安装Jenkins插件	327
	10.1.2 Jenkins插件全局配置管理	328
10.2	Jenkins持续集成基础配置	329
	10.2.1 新建项目	329
	10.2.2 构建项目	334

第11章 Appium常见问题处理方式 336

11.1	输入中文	337
11.2	滑动操作	337
11.3	滚动操作	339
11.4	输入Android按键	340
11.5	处理Popup Window	341
11.6	处理Toast	342
11.7	处理长按	345
11.8	处理下拉列表框	346
11.9	处理缩放	348
11.10	检查元素文本是否可见	348
11.11	启动其他App	350
11.12	并行测试	351
11.13	处理拖动	358
11.14	处理截图	359
11.15	隐式等待	362
11.16	显示等待方法	365

11.17	在编程中处理 adb 命令 ……… 366	11.19	区分 RemoteWebDriver、AppiumDriver、AndroidDriver 和 iOSDriver ………………… 368
11.18	区分 WebElement、MobileElement、AndroidElement 和 iOSElement ………………… 367	11.20	在代码中启动服务器 ………… 368
		11.21	PageFactory 注解 ……………… 371

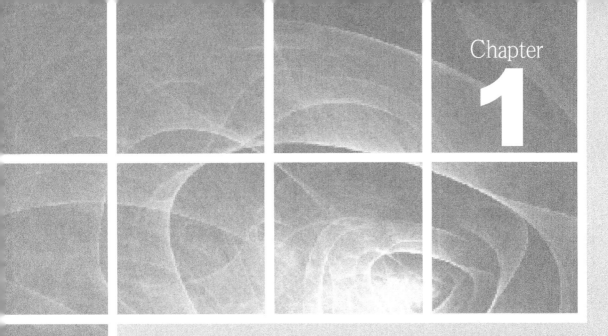

第 1 章

了解 App 测试

1.1　App 测试与普通软件测试的差异

1．软件载体的不同

传统软件都部署和安装在计算机（台式机和笔记本电脑）上，而 App 的载体是手机等智能移动终端，因此我们可以将手机这个概念扩充为"智能移动终端"或者"智能终端"。

2．软件规模的不同

传统软件的规模动辄几千万行代码，研发周期长达一两年，甚至数年之久。

App 软件很少有这么大的代码量，并且其研发周期也都是采用快速迭代、小步快跑的形式。其研发流程的快速迭代化、需求变更的速度，都是传统软件研发没有过的情形。

这给从传统测试转向 App 或手机测试的广大同仁带来了不小的挑战，其实也在挑战着传统的研发管理模式。

3．测试场景的不同

对于传统的 PC 端测试场景，读者都很熟悉，通过测试工具，并能够借用鼠标、键盘等外设进行操作，PC 的大屏幕实时显示各种测试数据，加之 PC 的性能也比较高，通过 PC 计算可以非常直观地产生各种测试图表，还能找到一点"测试的感觉"。转移到手机端测试后，很多人拿着手机，坐在一个角落里，面对一个操作有限、带有子屏幕的设备，感觉无从下手，甚至测试手机的过程貌似就是在"玩手机"。

面对新的测试场景，尤其是手机自动化测试，从技术上看，其难度和挑战不低于传统软件的自动化测试。新入行的同仁需要认真对待，转行 App 测试的同事需要积极转变心态，手机载体里面还是大有乾坤的。

4．测试关注点的不同

1）性能

在传统软件中，软件的性能可能聚焦在大数据情况下的查询效率、吞吐量等重量级指标上，而 App 的性能不仅包含传统的服务端性能指标，还包括客户端应用启动时间、应用安装后占用的磁盘空间大小、滑屏的响应时间。这些指标都会受到关注，而且是重点测试内容。

2）易用性

虽然传统软件测试也注重易用性，但是这个易用性测试在 App 软件测试中的地位无疑被提升到了很高。同质化的软件很多，用户将会选择最好用的软件。

3）吸引性

过去大部分时间，软件的吸引性只是挂在测试工程师的嘴边，甚至只被看作在"软件质量模型"（ISO9126 的易用性大项中的吸引性指标）中的一个名词。在 App 测试中，吸引性之重要性不言而喻。我们在很多时候把易用性当作了吸引性，其实吸引性的范畴更广泛。它是一个综合性概念，这项质量指标也在 App 测试中立刻被重视了起来。

4）App 的稳定性

在 App 时代，软件稳定性问题非常突出。当手机充当了我们生活、工作甚至社交等方面的"钥匙"时，软件的稳定性不仅仅给客户带来了一个使用感受方面的问题，更多的是关乎安全性的问题。比如一个支付软件频繁闪退、频繁死机，这样的软件还敢用吗？

5）自动化测试

虽然传统软件的自动化测试已经开展了很多年，而且业界各大公司也有自己专职的自动化开发团队、自动化执行团队，并且都各有建树。但是在 App 的开发节奏中，这种快速迭代、小巧灵活的软件形式，让 App 测试超乎以前所有的软件形式而存在。大家普遍认为，要做好 App 测试，必须要进行 App 自动化测试。但是 App 自动化测试又在快速迭代的软件开发周期中显得捉襟见肘，难度非常大，不仅考验着自动化测试手段，还考验着自动化测试的管理。

不管怎么困难，很多的测试类型和测试内容是离不开自动化测试的，离开自动化手段，则根本无法完成该测试内容，相关内容在第 2 章中会详述。但是手机载体的种种限制，使得 App 火爆的年代中，成熟的商用自动化测试工具方面形成了一个大大的空档期。所有测试界从业者，不得不挽起袖子，自力更生，在诸多开源框架的基础上，自己开发自动化测试框架，完成相关测试。这也是本书的主旨：在商业成熟的自动化测试工具不具备的情况下，帮助更多的测试团队开发相关自动化测试框架。

1.2　App 测试的难点

App 的测试难点确实很受关注，尤其是刚刚进入 App 测试领域的人员。根据笔者的相关经验以及和 App 专业测试人员的交流，App 的难点大概有以下 4 个方面：

- App 的兼容性测试；
- App 的稳定性测试；
- App 的功耗测试；
- App 的自动化测试。

我们可以深入地分析一下，在更大层面上，这些测试之所以让广大 App 测试同行感觉头疼，本质上是因为 App 自动化测试手段的不成熟。因为兼容性（详细的兼容性测试内容参见 2.3 节）、稳定性和功耗测试在很大程度上都要依赖自动化测试手段，才能高效和高质量完成。尤其是对于 App 的稳定性测试，自动化测试手段更是至关重要。

此外，兼容性测试在测试物料准备方面也存在困难，目前快速解决兼容性测试的突破口就是"云测试"，也就是一种在云概念下的自动化测试，但是目前的云测还处于起步阶段，测试深度和广度还有待加强。不过这已经是一个使用自动化测试手段来解决手机兼容性测试问题的好方法了。

对于稳定性测试，可以套用一个概念——MTBF 测试，这个测试要严重依赖自动化测试手段。功耗测试更是必须借助自动化测试手段来完成，为了使结果更为客观有效，甚至要用到机械臂这种"硬"的自动化工具，不仅仅是传统的"软"自动化测试方案了。

1.3　App 测试中的网络信号概述

目前我国的移动通信网络信号主要是 2G、3G、4G 的混合信号。2G 信号的信号制式比较简单，就是 GSM。3G、4G 信号就比较复杂，简要描述见表 1-1 和表 1-2。

一般来讲，专门从事手机终端测试的工作人员，需要重点关注网络信号，而专门从事 App 测试的工作人员则不太需要关注移动网络信号。

表 1-1 国内 3G 信号

技术标准	TD-SCDMA	WCDMA	CDMA2000
上行速率	2.8Mbit/s	5.76Mbit/s	1.8Mbit/s
下行速率	3.6Mbit/s	14.4Mbit/s	3.1Mbit/s
运营商	中国移动	中国联通	中国电信
备注	中国自主技术	全球推广	技术优秀，产业链一家独大

表 1-2 国内 4G 信号

技术标准	TD-LTE	FDD-LTE
上行速率	50Mbit/s	40Mbit/s
下行速率	100Mbit/s	150Mbit/s
运营商	中国移动、中国联通、中国电信	中国联通、中国电信
备注	中国自主技术	中国联通、中国电信 4G 均为混合组网

但是，CSFB（Circuit Switched FallBack，电路域回落）的概念也需要了解一下，以便于在一些冲突测试的场景中合理选择冲突时间。

在 4G 网络时代，不管是 TD-LTE 还是 FDD-LTE，当进行语音业务时，都要将网络回落到 2G 网络上进行语音通信，待通话完成后，再返回 4G 网络。这是对 CSFB 机制一个最简单的描述。

其实每个手机都存在一个 CSFB，或者返回时的延时问题、返回驻留 4G 网是否成功的问题。如果一个手机正在使用某 App 时，而另一个手机拨电话进来，这时候 4G 网络就会回落到 2G 网络，4G 网络中断；当重回 4G 网络后，App 的唤醒表现等，都是需要特别注意的情况。

1.4　智能终端中的 App 测试

智能终端的外延很广，手机、PAD 是常见的智能化终端产品。其实智能可穿戴设备（比如手环、手表）也是智能终端。家里的智能路由器，甚至一些机顶盒在广义上都可以纳入智能终端行列。

国内的手机品牌很多，经历了功能机、智能机这样一个发展历程。近几年功能机逐渐退出历史舞台，智能机占有了绝对的优势，并且智能机已经不仅是一个通信工具，还成为人们生活的中心支撑点。手机的互联互通领域越来越大，人们的生活和工作对手机的依赖程度也越来越大。与其说是对手机的依赖，不如说是对手机上服务的依赖，即对数量庞大、涵盖领域广泛的 App 的依赖。App 的蓬勃发展依赖于手机这个载体在硬件配置上的提升，手机深度嵌入人们的生活和工作，且需要靠 App 提供良好的内容和服务。两者相辅相成，缺一不可。

传统的国产品牌中华酷联，在进入 2017 年时就已经发生了很大的变化。智能手机行业的发展也是日新月异，不断地涌现出新的品牌和具备新功能的智能手机。每一次智能手机功能的提升，都会给 App 测试带来机遇和挑战。

对于在手机终端上做 App 测试来讲，常见的关注点有以下几个方面。

1．移动终端品牌

在国内不仅要注意众多的手机品牌，测试时要考虑尽可能多地覆盖比较流行的品牌，同时还要考虑运营商这个维度，即终端还可以按照运营商的定制与否划分为通用版机、移动定制机、联通定制机、电信定制机。

当然，随着 4G、未来 5G 技术的普及和国家要求的全频段支持，运营商定制机的特殊化越来越淡了。各厂商生产的手机都将以通用性版本为主。

2．手机的解决方案提供商

目前的厂商主要有高通、华为海思、MTK、大唐联芯、马维尔、展讯科技等。

3．Android 的开源特性导致的定制系统差异化

操作系统繁杂，在智能手机兴起的时候，苹果的 iOS、谷歌的 Android、诺基亚的塞班、微软的 Windows Mobile 和黑莓的黑莓系统并存。

经过一段时间的洗礼，现在的智能手机系统主要还剩 3 家——iOS、Android 和 Windows Mobile。

iOS 和 Windows Mobile 厂商比较单一，尤其是 iOS 厂商，只有苹果公司一家。而 Android 系统的手机遍地开花，但仅 Android 系统就给 App 测试带来了很大的挑战，同一版本的 Android 系统在各个手机品牌厂商的定制化中皆有不同。另外，设备品牌纷杂，而且市场混乱，Android 手机不仅有正规品牌厂商提供不同版本的大量手机，还有少量的山寨机。

Chapter 2

第 2 章

App 测试类型

本章主要讲述 App 测试类型中传统测试容易忽略的部分，这部分内容仅是笔者多年来从事移动终端测试及 App 测试时的一些经验积累，供读者参考，希望给读者一个启发，引发更多的共鸣和思维亮点。

App 及手机测试过程中，有很多是传统测试所没有的，或者是不太重视的，如网络兼容性、SIM 卡兼容性、多部手机间的冲突测试、GPS、内置摄像头、手机丢失后的安全防御等。下面按照测试类型逐一讲述相关测试的内容重点。

2.1 功能测试

功能测试，通常的定义就是测试功能的可执行性和有效性。

以下内容没有覆盖到功能测试的所有方面，读者都很熟悉的常规内容就不再讲述了。在 App 功能测试中，有一些传统软件测试里不太常见的关注点，以下权当抛砖引玉，启发一下读者在 App 功能测试中的思考维度。

2.1.1 高级别事件响应

高级别事件响应也可以归为冲突测试的一个类别。这里单独提出来进行介绍，能让读者更准确地理解冲突测试，从而使设计思路更清晰。

比如，当用户正在操作一个 App 时，闹铃响了，这里的闹铃显然比该 App 相关操作的事件级别要高，因为即使在关机时，闹铃也照样会响，在不主动干预的情况下，这个事件是不可阻止的。同理，我们也可以把其他 App 定期产生的推送消息当作一种高级别事件，拿到测试场景中来进行设计。当然，当 App 自动化测试的环境初始化时，一定要阻止这些事件响应的发生，应该在手机的相关设置里将其屏蔽掉。否则，这肯定会影响 App 自动化测试程序的正常运行。

2.1.2 第三方应用打断

广义上，上述高级别事件响应也属于一种第三方应用打断场景，但是这里归纳出来的第三方应用打断，重点考虑的是多终端场景。比如，A 手机正在操作一个 App，B 手机给 A 打电话，C 手机给 A 手机发短信，D 手机给 A 手机发送邮件。当然，还可以根据场景扩展到更多的第三方终端，让他们来发送 QQ 消息、微信消息，还有手机上其他应用产生的推送消息等。关于这部分测试，使用自动化测试手段才能化繁为简，并且取得比手工测试更准确、更客观的测试结果。自动化测试手段能够编写同一时钟下的相关操作，以确保测试的及时性和准确性。而确保动作序列的流程、最大限度地提高容错性和实现相关的等待时延判断，是这种自动化测试程序的关键所在。

2.1.3　通信录的备份恢复功能

测试人员需要充分考虑新手机开机时的备份恢复功能，刷机前后的相关备份恢复功能、增量备份恢复、全量备份恢复、备份恢复时的异常情况测试。这些是很多客户都关注的功能，不管是手机本身，还是相关 App，如果能够灵活、准确、高效地提供此项功能，那么在特定场景下的用户满意度将会非常高。

2.1.4　手机和其他外设产品的互联互通

目前，智能手机给人们提供的服务已经远远超过电话、短信业务，也不仅仅是手机上安装的相关 App 提供的服务。手机及某些 App 和其他外部设备的互联互通极其常见，而且很多时候也是非常必要的。比如与蓝牙音箱连接，手机可外部播放音乐；与智能电视连接，手机甚至可以用来做遥控器；与小区的门禁系统连接，手机就是门禁卡。此外，还可以与汽车影音系统和智能可穿戴设备连接，实现更多功能。手机本身具备的功能，或者手机上某款 App 具备的功能，外部设备的互联互通肯定不能不进行测试。连接方式也有很多，有通过电信网连接的，有通过蓝牙连接的，有通过 Wi-Fi 连接的，也许还可以用 ZigBee、GPS 等连接，不一而足。我们在选择测试场景时可以参考一下。

2.2　稳定性测试

传统硬件测试中的可靠性测试，在软件测试中通常叫作稳定性测试。软件稳定性的测试方法借鉴硬件的可靠性测试非常多，目前广泛应用的硬件可靠性指标有 MTTF、MTTR 和 MTBF 三个指标，较为通用权威的标准是 MIL-HDBK-217、IEC 61508 和 Bellcore，分别是美国国防部、AT&T 贝尔实验室和国际电工委员会提出的标准。下面介绍一下 MTTF、MTTR 和 MTBF 三个指标。

- MTTF（Mean Time To Failure，平均失效前时间）。平均失效前时间是指系统/产品平均正常运行多长时间，才发生一次故障。我们也可以称它为平均无故障时间。这个指标值越大越好，最好永远不会发生故障。
- MTTR（Mean Time To Restoration，平均修复时间）。平均修复时间就是修复产品所用的平均时间，即从出现故障到修复故障的这段时间。在统计时，这个时间要包括获取到产品的时间、维修团队的响应时间、记录所有任务的时间，还有将产品重新投入使用的时间。当然，这个指标越小越好，出现故障后最好能立刻修复。
- MTBF（Mean Time Between Failures，平均故障间隔时间）。平均故障间隔时间，是指修复产品两次相邻故障之间的平均时间。这个指标对于经常做稳定性测试的人员耳熟能详，它是体现产品持续正常工作多长时间的一种能力。这个指标的值也是越大越好。

通过以上三个指标的概念可以看出，它们之间存在着这样的公式，即 MTBF = MTTF + MTTR。在实际测试度量中可以发现，MTTR 的值远远小于 MTTF，所以一般直接用 MTBF 值表示系统/产品的稳定性。

更专业的硬件或电子元器件的可靠性测试指标是如何度量或计算的，这里就不做介绍了，下面说一下软件测试中怎么使用 MTBF 值。一般我们看到某款计算机或者电子产品在宣传中称，该产品经过了几千小时、几万小时的稳定性测试，难道该产品真的是单一产品连续测试几千小时吗？1 年按 365 天算，几万小时就是好几年，如果一个产品测试就要几年，那么电子产品就永远无法上市。

软件的稳定性测试可以借鉴以下公式：

MTBF（时间/次）=N 个选样产品总运行时间之和/N 个产品发现的指标 BUG 之和

也就是说，当测试 MTBF 时，可以选择多个产品并行测试，将其并行运行时间和期间发现的 BUG 数量进行累加，这样测出几千、几万小时的指标值就不需要那么长的实际时间。值得注意的是，这里所说的指标 BUG，不完全等同于功能测试时找到的一般性 BUG（也就是说，稳定性测试中的指标 BUG）。我们可以界定几类重点关注的 BUG，而把一些不是很重要的 BUG 忽略不计。比如手机的稳定性测试中，我们可以只关心闪退、花屏、黑屏、死机等 BUG，而出现的其他 BUG 可以不计入 BUG 数量，这样可以让 MTBF 这个值更能表现出产品稳定性的特定意义。当然，把所有出现的 BUG 都计入指标 BUG 也是可以的。

传统的软件稳定性测试还有一种要求，只要发现后台进程 core dump，或修改 BUG 导致后台进行重启，稳定性测试就重新计时，即不管指标 BUG 怎么界定，后台进程只要挂掉一次，稳定性要从头再做，时间不可累计。

至于手机测试和 App 测试中的稳定性需要测试多久，还没有像硬件产品那样比较丰富的参考对象和相应的标准。关于手机的稳定性测试，一些手机厂商曾给出过一个参考指标是几百小时这样的量级。当然，不管是多久，对于一款 App 最少要测试 24 小时的稳定性，即使是这样，进行 24 小时连续不间断的手工测试也很难做到，如果要进行 $N×24$ 小时的稳定性测试，那必须借助自动化手段来完成。所以自动化测试手段在手机和 App 的稳定性测试中是一个必选途径。

2.3 兼容性测试

兼容性测试本身比较复杂，实施难度也很大，历来都被测试界公认为"又脏又累"的工作。下面没有完全展开兼容性测试分析，仅仅给出 App 或手机测试中与兼容性相关的常见思考维度，以供参考。

2.3.1 手机品牌

Android 的开源特性，使得同一个大的 Android 版本在不同的厂商进行的定制化不同，且操作系统千差万别。在我们无法统计和分析到底有哪些不同的时候，我们要尽可能多地兼容手机的品牌，以及相同品牌下不同型号的手机。

业界的有些实践经验就是重点要兼容 3 个季度内的手机，我们可以权且把它们叫作"新机"，上市在一年或一年半左右的机型，可以称为"主流机型"。根据公司的实力和渠道，要尽可能多地覆盖测试这些品牌和机型。对于时间更久的机型，可以适当挑选典型的机型进行测试，覆盖太全可能导致测试成本太高，测试效果未必能得到正向收益。

2.3.2 硬件种类

常用的智能硬件有以下几种。

- 智能手机。按操作系统来划分，有 iOS、Android、Window Mobile 等智能手机。
- PAD。App 还要考虑各种操作系统上的 PAD 产品，因为 PAD 并不属于通信设备，在硬件构造上和手机不同，屏幕尺寸也比手机大很多。对于一些平板类笔记本电脑，这种二合一的结合体也是一个新生事物，要酌情加以考虑。
- 智能可穿戴设备。常见的智能可穿戴设备有智能手表、智能手环。目前的硬件产品上所承载的 App 可能并不多，这也受限于这类产品的硬件能力和使用场景，但是未来这类产品肯定会被广泛应用，并且所承载的 App 会越来越多。
- 车载终端。对于车载导航、车载的语音娱乐系统平台等车载终端，App 的承载需求也很突出。
- 智能电视/智能机顶盒。虽然智能电视/机顶盒上大部分都是影音类播放 App，但是随着智能电视的普及，它在很多场合可以作为智慧家庭的重要入口点，需要的 App 种类和数量也是可以期待的。

随着技术的发展，可能日后还会出现更多的智能硬件品种，这里列举如上种类，旨在扩宽我们测试选型的角度。对某些 App 产品，如果应用领域本身就很宽，测试载体就不可仅仅局限在智能手机范畴，也可以在移动互联网、物联网的领域多多寻找，这样所测试的 App 产品可能会有更广阔的应用领域。

2.3.3 芯片种类

对于 App 测试来讲，芯片种类并不是必须要兼容的内容，这部分内容过于底层。不过对于移动终端产品来讲，不同芯片的解决方案不同，产品是要重点进行测试的。为了保持知识体系的完整性，将这部分内容也归纳进来，对于 App 测试者来讲，可以进行了解参考。

目前，主要的芯片厂商有美国的高通、美国的苹果、中国深圳的华为海思、中国上海的展讯科技、中国上海的大唐联芯、中国台湾的 MTK（联发科）、韩国的三星、美国 Marvell（马维尔）等。在 TD-LTE、FDD-LTE、TD-SCDMA、WCDMA 等领域，各家都有各家的长处。我国市面上手机产品使用的芯片解决方案几乎都是以上厂家提供的，芯片质量的好坏直接决定了手机的各种质量指标。所以，有的时候，同样一个产品在这款手机上稳定性很高，在另一款手机上稳定性不高，这是因为手机使用的芯片不同造成的。

2.3.4 分辨率

分辨率简言之就是屏幕的精密度,即一个屏幕上容纳像素点的多少的衡量。分辨率越高,我们在同一屏幕上所看到的图像越清晰。比如,当分辨率较高时,屏幕上一行能显示 80 个汉字;当分辨率较低时,屏幕上一行只能看到 40 个汉字。

对于这项兼容性,不仅 App 要测试,传统软件也要测试。兼容性就是测试软件对分辨率的自适应性,即会不会因为分辨率改变界面显示情况。

智能终端的分辨率表述和 PC 的分辨率表述是一致的,都使用行×列像素的表示方法,如常见的 540×960 像素、640×1136 像素、720×1280 像素、800×1280 像素、1024×768 像素、2048×1536 像素等。

选择测试载体时可以按照分辨率来选择。当然,更简单的选择标准,可以按照屏幕尺寸来选择,比如 4.3 英寸屏、5 英寸屏、5.5 英寸屏、5.8 英寸屏等。这时候需要注意的是,同样尺寸的屏幕分辨率未必一样。所以在选择时,统筹考虑是有必要的。在测试中,最常见的就是对手机屏幕进行旋转,可能会发生很多类型的错误。

2.3.5 各种无线网络的兼容性

针对各种无线网络的兼容性,App 测试可以选择性进行覆盖,因此智能终端测试就必须完成。

移动通信网络信号在 1.3 节中已经介绍过,这里不再详述。以下再简要归纳这部分的选择范围:各种通信网络连接、Wi-Fi 网络、蓝牙、GPS 等常见的无线连接在兼容性测试中需要考虑。

2.3.6 第三方软件兼容性

第三方软件兼容性测试主要用于测试 App 产品与本机预装的 App 及主流 App 是否兼容。

预装 App,就是每部手机出厂时,厂家或者相应的销售渠道代理给手机预装的一些 App。目前,很多预装软件是在用户使用模式下无法卸载的,即使不使用,这部分 App 也卸载不掉。除非进入手机的工厂模式下才能将其卸载。

主流 App 目前也没有标准的定义，我们可以根据国内几大 App Store 中 App 的下载量排名来选择，也可以根据自己的经验来进行判断。

别外，和被测 App 属于同行竞争产品的 App，以及和被测软件有交互操作的 App 也需要重点测试。

2.4 性能测试

App 的性能测试非常重要，也是 App 测试中频率最高的必测内容。性能测试简单来讲就是，评估典型应用场景下 App 产品对系统资源的使用情况。这里的典型应用场景，一定要根据用户实际使用场景、软件极限应用场景、软件需求规格说明书（SRS）的相关标准来综合考虑，不同场景下的性能测试效果会差别很大，同时对软件质量的保障力度也不尽相同。

传统性能测试从大的方面讲主要测试两个方向的特性，一个是空间特性，另一个就是时间特性。在 App 性能测试中，功耗测试（也叫电量测试）经常被划分到性能测试中。常见的性能测试评估指标有 CPU 占用率、内存占用率、上下行流量测试、耗时、流畅度、电量。

具体 App 的性能自动化测试不是本书的重点，想深入了解相关内容请读者参阅相关专业书籍。

2.5 网络测试

2.5.1 室内网络测试

室内网络测试就是在室内固定地点，选择移动网络较好或者较差的地点，自行设计网络信号强弱点，还可以在室内连接稳定的 Wi-Fi、蓝牙等无线网络进行相关测试。

2.5.2 外网测试

外网测试包含常说的路测、户外拨测。

外网测试情况比较复杂。对于手机的外网测试，国家规定的有固定的城市，每个城市里有固定的线路，并且各运营商有固定的测试用例。此外，对于手机的基站信号测试，还要考虑高铁、车站、旷野等场景进行测试。

App 的外网测试可以不沿用手机路测那么严格的路线，但是高铁、公路、郊区、市区、超市、车站、地下车库等场景还是要考虑的。

外网测试主要模拟客户使用的网络环境，如 Wi-Fi、GPS、北斗，以及移动、联通、电信的 2G、3G、4G 网络。

场景包括在高山、丘陵、火车、高铁等特殊环境下的 App 测试。

信号被屏蔽后，也要进行 App 的测试。

2.5.3 弱场测试

弱场测试一般是指在信号比较弱的场所进行的测试。

一般在测试时，选择地下车库、地下室、地铁上、地铁下层换乘厅、地下超市、电梯等场所即可以满足日常弱场测试场景。虽然很多实践中也把弱场测试划分到外场测试里面，但是最好还是提取出来单独设计。这个弱场测试对手机测试来讲更为重要一些。

2.6 异常测试

异常测试包括以下几种情况下的测试。

（1）各种网络信号的网络中断异常。

（2）SIM 卡松动，这可以采取 SIM 卡插拔手段模拟实现。

（3）低电量。

（4）手机内存占或 CPU 占用率达 100%。

（5）手机死机或卡死。

对于后面两种情况，可以自行研发一个 App。其主要功能是快速占满内存和 CPU，或让进程产生异常，使手机快速产生卡死现象，因为手工完成这些操作会很麻烦。

2.7 发布测试

要发布测试,可以按以下步骤进行。

发布测试包括以下内容。

(1) 检查安装包大小。

(2) 检查版本号、语言。

(3) 安装和反安装测试。

(4) 用其他辅助工具(如 91 助手、豆瓣荚等)安装、卸载测试。

(5) 在线升级测试,相近版本及跨版本升级。

(6) 验证数字签名。

2.8 用户界面测试

用户界面测试旨在测试用户界面(如菜单、对话框、窗口和其他可见控件布局、风格)是否满足客户要求,文字是否正确,页面是否美观、完整,文字、图片组合是否完美,操作是否友好等。

界面测试的目标是确保用户界面会通过测试对象的功能来为用户提供相应的访问或浏览功能,确保用户界面符合公司或行业的标准,包括易理解性、易操作性、易学习性等测试点。

下面重点介绍图形测试和内容测试。

2.8.1 图形测试

图形测试包括以下内容。

(1) 横向或纵向比较,确认各控件操作方式统一。

(2) 自适应界面设计,内容根据分辨率大小自适应。

(3) 测试页面标签风格是否统一。

(4) 测试页面是否美观。

(5) 页面的图片应有其实际意义，而且要求整体有序美观。

(6) 图片质量高且图片尺寸在符合设计要求的情况下应尽量小。

(7) 界面整体使用的颜色不宜过多。

2.8.2　内容测试

内容测试包括以下内容。

(1) 测试输入框说明文字的内容与系统功能是否一致。

(2) 测试文字长度是否加以限制。

(3) 测试文字内容是否表意不明。

(4) 测试是否有错别字。

(5) 测试信息是否有中英夹杂或中文中夹杂其他语言的情况，如果有，则需要核对需求规格说明书，或找前台开发负责人进行确认。通常，中外文夹杂就是 BUG。

(6) 测试是否有公司、行业或法律法规所规定的敏感性词汇。

(7) 测试图的合法性，如是否涉及版权、专利、隐私等问题。

2.9　冲突测试

　　2.1.2 节提到的第三方应用打断的场景，其实也就是冲突测试的内容。此部分可以放在功能测试里统一考量，也可以单独提取出来作为一个测试项目。因为人们经常会提到冲突测试，所以在此单列一节，但是主要内容在 2.1.1 节里已经详细描述了。下面给出几个移动终端上 App 冲突测试的典型场景实例，有助于读者拓展思路。

2.9.1　按键打断

　　考虑到手机下方功能键有三个键的情况，还有一个键的情况，根据场景，可以设计不同的按键进行干扰打断。

另外，还要考虑到关机/锁屏键的干扰打断，以及其他手机上的按键功能。

2.9.2 程序后台相互切换

在 App 之间频繁切换的场景很多，尤其是多个交互的 App 之间的业务协作切换（如 12306 订票和支付宝之间的切换），需要重点考虑。

2.9.3 网络切换

网络切换包括：两种 4G 网络之间的切换，三家运营商之间的网络切换，移动网络和 Wi-Fi 网络之间的相互切换。使用正交实验方法进行用例组合，会实现比较完整的测试。

2.9.4 待机唤醒

在手机进入待机状态之后，对于 App 要设计几个待机时长的等价类。至于待机的具体时长，可以通过咨询相应的开发人员，了解一下该 App 前台失效的等待时间阈值，在阈值边界处进行边界值（上点和离点）选取。

2.10 接口测试

App 的接口测试也是实际测试中使用非常频繁的一种测试类型。尤其是在快速迭代、一些紧急补丁的场景下，设计好充分的接口测试用例，自动进行快速测试比传统界面级别的功能测试的效率高很多。当然，效果也是很不错的。

服务器端一般会提供 JSON（JSON 语法是 JavaScript 语法的子集）格式的数据给客户端，这种格式就是键值对，如 "Name" : "David"。

在服务器端需要进行接口测试，确保服务器端提供的接口和转换的 JSON 内容正确，对分支、异常流有相应的返回值。

此部分测试可以采用 ITest 框架完成，最方便的方法是采用 HttpClient。

下面给出几个接口测试的接口划分场景，以供参考。

（1）客户端和服务器端交互测试。

（2）测试客户端的数据更新和服务器端数据是否一致。

（3）当更新客户端时，客户端和服务器端断开。

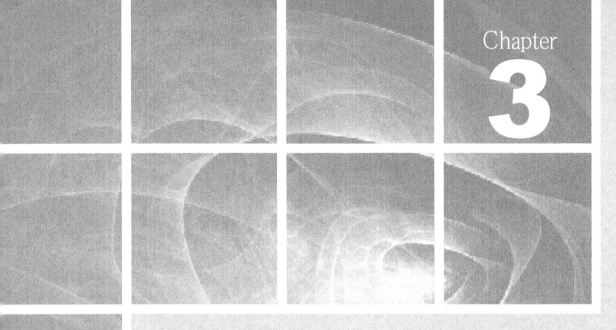

第 3 章

Java 编程环境构建

本书所介绍的 Appium 是基于 Java 语言实现的，学会配置环境是编程以及使用 Appium 进行自动化测试的基础。在进行测试脚本开发前，最重要的步骤是获取 JDK 并如何正确地安装 JDK，正确地进行 Eclipse 的安装和配置，之后才能继续学习 Java 和编写代码。

Maven 是 Java 编程中一个重要的项目管理工具。如果每个 Java 项目的目录结构没有统一的标准，配置文件到处都是，Jar 包需要一个个去下载，然后导入到项目中，这就会非常麻烦，因此就要用到 Maven。Maven 有助于统一管理 Jar 包。

本章主要介绍 JDK/Eclipse 的安装和配置。

3.1　安装 JDK 与配置环境变量

Android 应用程序是用 Java 语言编写的，所以首先必须安装 JDK。而为了搭建 Eclipse 的 Android 开发环境，也需要先安装 JDK，所以本章先介绍 JDK 的安装和相关配置。

3.1.1　下载 JDK

在 Oracle 官方网站，可以下载 JDK。在图 3-1 所示界面中，选择 JDK DOWNLOAD 按钮，即可下载 JDK。

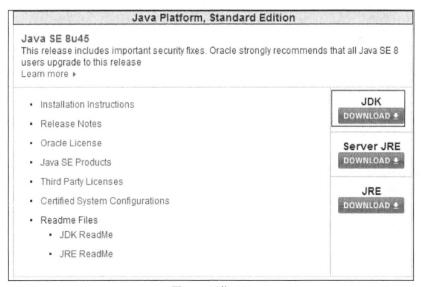

图 3-1　下载 JDK

进入版本选择页面后，根据操作系统，选择合适的 JDK 版本进行下载。笔者所使用系统为 Windows 7 的 64 位操作系统，所以选择下载 jdk-8u45-windows-x64.exe，如图 3-2 所示。

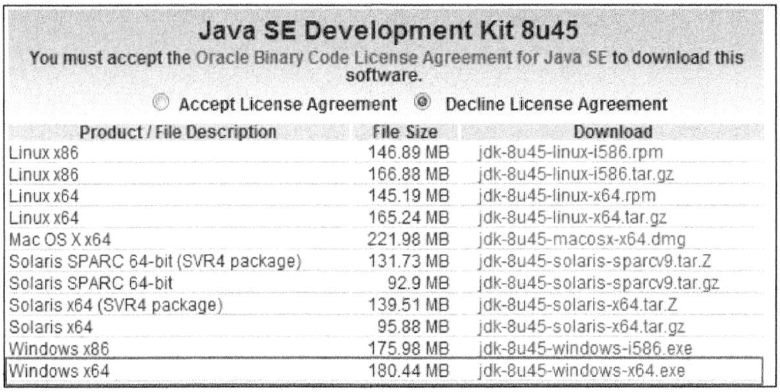

Java SE Development Kit 8u45

You must accept the Oracle Binary Code License Agreement for Java SE to download this software.

Accept License Agreement　Decline License Agreement

Product / File Description	File Size	Download
Linux x86	146.89 MB	jdk-8u45-linux-i586.rpm
Linux x86	166.88 MB	jdk-8u45-linux-i586.tar.gz
Linux x64	145.19 MB	jdk-8u45-linux-x64.rpm
Linux x64	165.24 MB	jdk-8u45-linux-x64.tar.gz
Mac OS X x64	221.98 MB	jdk-8u45-macosx-x64.dmg
Solaris SPARC 64-bit (SVR4 package)	131.73 MB	jdk-8u45-solaris-sparcv9.tar.Z
Solaris SPARC 64-bit	92.9 MB	jdk-8u45-solaris-sparcv9.tar.gz
Solaris x64 (SVR4 package)	139.51 MB	jdk-8u45-solaris-x64.tar.Z
Solaris x64	95.88 MB	jdk-8u45-solaris-x64.tar.gz
Windows x86	175.98 MB	jdk-8u45-windows-i586.exe
Windows x64	180.44 MB	jdk-8u45-windows-x64.exe

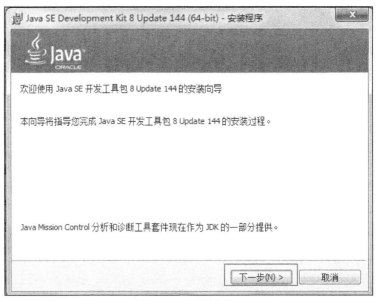

图 3-2　JDK 版本选择

3.1.2　安装 JDK

双击已下载的 JDK 安装文件 jdk-8u144-windows-x64.exe，将出现如图 3-3 所示的界面。

图 3-3　JDK 安装向导 1

单击“下一步”按钮，配置安装目录，这里选择默认安装目录，如图 3-4 所示。

图 3-4　JDK 安装向导 2

开始安装 JDK 相关文件，如图 3-5 所示。

图 3-5　JDK 安装向导 3

待相关文件安装完成后，出现如图 3-6 所示界面，提示 JDK 已经安装成功。

图 3-6　JDK 安装完成

3.1.3　配置环境变量

环境变量在操作系统中是一个具有特定名字的对象。它包含了一个或者多个应用程序将使用到的信息。例如 Windows 中的 Path 环境变量，当要求系统运行一个程序而没有告诉它程序所在的完整路径时，除了在当前目录下面寻找此程序外，系统还应到 Path 中指定的路径去寻找。用户通过设置环境变量，来更方便地运行进程。

1．配置环境变量

JDK 的安装需要配置 3 个环境变量，分别是 JAVA_HOME、Path 和 CLASSPATH。

Windows 7 系统下，右击"计算机"，选择"属性"→"高级系统设置"→"环境变量"，弹出"环境变量"对话框，如图 3-7 所示。在"系统变量"下，单击"新建"按钮，弹出"新建系统变量"对话框。

图 3-7　环境变量的位置

图 3-8　新建环境变量

2. 配置 JAVA_HOME 环境变量

如果设置 JAVA_HOME 目录的路径,那么以后引用这个路径时,只须输入"%JAVA_HOME%"即可避免每次引用都输入较长的路径字符串。在"环境变量"对话框中,单击"新建"按钮,弹出"新建系统变量"对话框如图3-8所示,在"变量名(N)"文本框中输入"JAVA_HOME",在"变量值(V)"文本框中输入JDK的安装目录。

因为笔者的JDK安装目录为"C:\Program Files\Java\jdk1.8.0_144",所以将此路径输入"变量值(V)"文本框内。操作结果如图3-9所示。

图 3-9 配置环境变量

3. 配置 Path 路径

选择"Path"系统环境变量,然后单击"编辑"按钮,在变量值最后面添加"%JAVA_HOME%\bin;"。注意,最后的分号";"不能缺少。这句话的含义是将JDK安装目录下的bin目录配置到Path环境变量中,因为JDK的安装目录已经配置到JAVA_HOME下,所以这里就可以使用%JAVA_HOME%的形式引用它,避免了二次书写。操作结果如图3-10所示。

图 3-10 配置 Path 路径

4. 配置 CLASSPATH

为了能够成功运行Java类,除了JAVA_HOME和Path之外,还需要配置Java的CLASSPATH。

在"环境变量"对话框,单击"新建"按钮,创建一个系统变量把"变量名(N)"设置为"CLASSPATH",把"变量值(V)"设置为"%JAVA_HOME%\lib\tools.jar;%JAVA_HOME%

\lib\dt.jar;"。配置过程如图 3-11 所示。

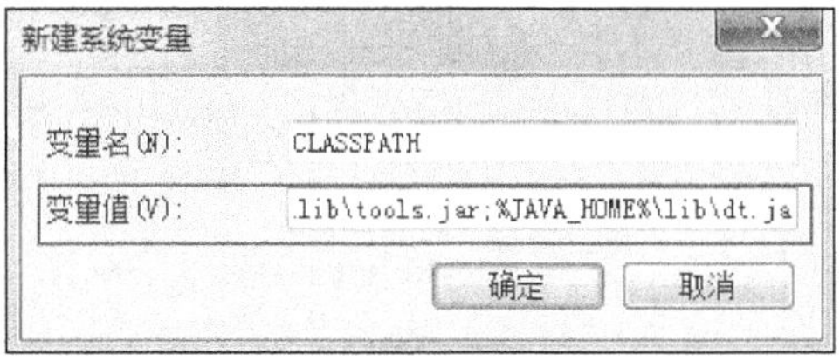

图 3-11　配置 CLASSPATH

5．验证 JDK 安装是否成功

添加环境变量后，用户可以打开 Windows 命令行窗口来验证。在图 3-12 所示"运行"对话框中输入"cmd"，单击"确定"按钮。

图 3-12　在命令行窗口中输入 cmd

执行命令"java -version"可以验证环境变量是否添加成功。如果添加成功，会显示出安装的 Java 版本，如图 3-13 所示。

图 3-13　验证 JDK 安装是否成功

如果未能成功出现 Java 版本信息，则说明环境变量配置错误，请仔细核对上一部分的配置内容。

3.2 安装与配置 Eclipse

用户可根据个人喜好选择对应的集成开发工具，如 Eclipse 是目前被广大 Java 开发者普遍使用的一款 IDE，产品易用性很高，使用受众广大，所以本书也基于 Eclipse 进行编写。推荐下载 Eclipse Luna Java EE 版本，它自带 Maven 插件，性能比较稳定。实际工作中使用该版本就不用再安装 Maven 插件，但是后面会介绍一下 Maven 插件的安装，以便在非自带 Maven 插件的 Eclipse 平台下开展工作。

3.2.1 安装 Eclipse

Eclipse 可从 Eclipse 官网下载。注意，根据提示下载 Windows、Mac OS 版本或 Linux 版本，并选择 32 位或者 64 位对应的安装软件，如图 3-14 所示。

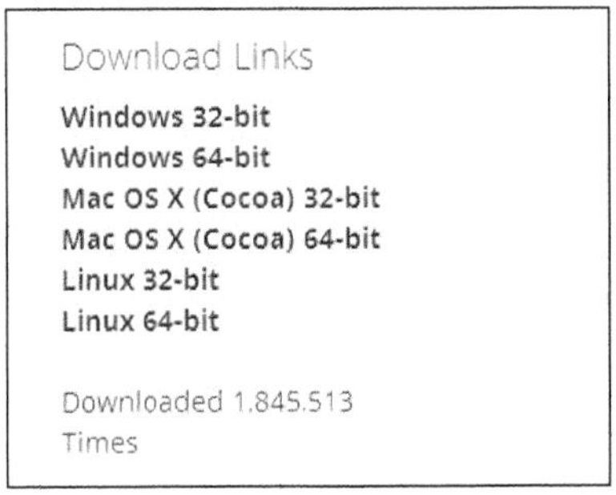

图 3-14　下载 Eclipse

下载后直接解压到任意目录，无须安装，双击解压缩目录下的"eclipse.exe"，便可安装 Eclipse。

 Android SDK 只支持 Eclipse 3.6 以上版本。

3.2.2 Eclipse 常用配置

在编码的过程中，经常会碰到一些关于 Eclipse 的常用配置问题，这里介绍两种。在 Eclipse 中进行如下配置，会让我们的工作变得更顺畅、更轻松。

1. 更改编码

以更改 Eclipse 的编码为 UTF-8 为例。操作如下。

打开 Eclipse，在菜单栏中选择 Windows→Preferences，在弹出的 Preferences 对话框中，在左侧面板中，选择 Text File Encoding，单击 Other 单选按钮，并从后面的下拉框中选择"UTF-8"，如图 3-15 所示。

图 3-15　更改编码格式

2. 更改 JDK 版本

每个 Eclipse 版本都带有默认的 JDK，但未必是项目所需要的 JDK 版本，所以可以手动将合适的 JDK 版本配置到 Eclipse 中。

打开 Eclipse，在菜单栏中选择 Windows→Preferences，在弹出的 Preferences 对话框中，在左侧面板中，选择 Java→Installed JREs，然后选择前面已安装 JDK 的路径，如图 3-16 所示。

图 3-16　更改 JDK 版本

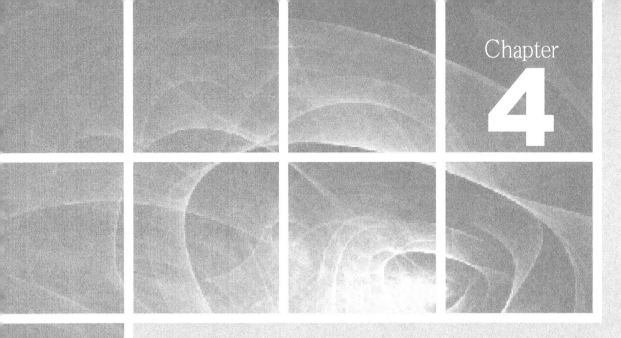

第 4 章

Java 语言基础

Java 是一门面向对象的编程语言，具有功能强大和简单易用两大特征。本书主要介绍 Android 环境下的测试，由于 Android 环境使用 Java 语言编写，因此掌握 Java 语言能够为以后持续性的平台集成打下基础，并且有助于快速学习其他语言。另外，我们通过学习 Java 还可以快速掌握面向对象程序设计的思想。

本章对 Java 的语法进行指导性的讲解，主要是为了配合自动化测试的相关内容，因此不可能很系统，假设读者具备一定的 Java 语言编码基础。如果学习本章有难度，建议专门找一本 Java 基础教程系统地学习 Java 语言。

4.1　Java 简介

通常，一个 Java 项目（Project）都包含零个到多个包（Package），每个包中包含许多 Java 类（Class），每个类中包含多个函数（Function），每个函数中包括一条条执行语句。在很多的编程语言中，函数是最小的可执行单元。

下面创建一个书店的程序。结合这个程序，展示和讲解自动化测试需要用到的知识。一步步按照书中的操作和代码进行学习，就可以掌握基本的 Java 编程知识。

4.2　第一个 Java 应用项目

首先，创建一个 Java 应用项目来学习 Java 语法知识。打开 Eclipse，在菜单栏中选择 File→New→Java Project，如图 4-1 所示。

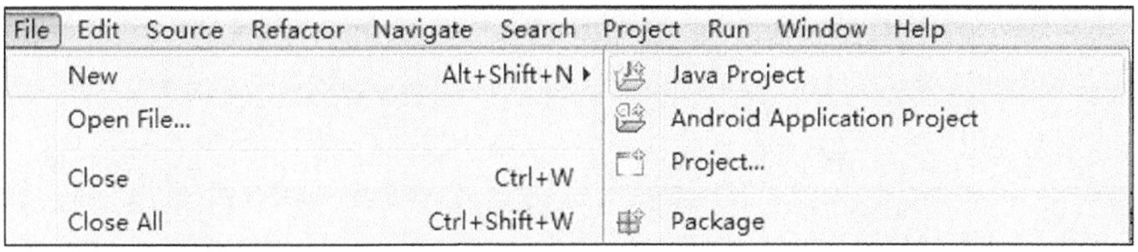

图 4-1　新建 Java 项目

在弹出的窗口里将 Project name 设置为 JavaCodeSample，如图 4-2 所示。

图 4-2　输入项目名

单击 Finish 按钮完成 Java 项目的创建。创建 Java 项目后的目录如图 4-3 所示。

图 4-3　Java 项目目录

4.3　函数

Java 项目新建成功后，列表中会有个 src 目录。这个目录一般用来存放源代码。在 src 目录上右击，从上下文菜单中选择 New→Class，创建一个类，如图 4-4 所示。

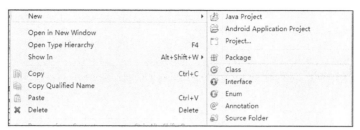

图 4-4 新建 Java 类

在弹出的 New Java Class 窗口输入类名和包名。类和包的概念分别会在 4.4 节和 4.5 节中详细讲述，这里先搭建示例环境。如果不输入包名，系统则会自动创建一个包，包名为 default package。这里不输入包名，输入的类名为 Book，如图 4-5 所示。

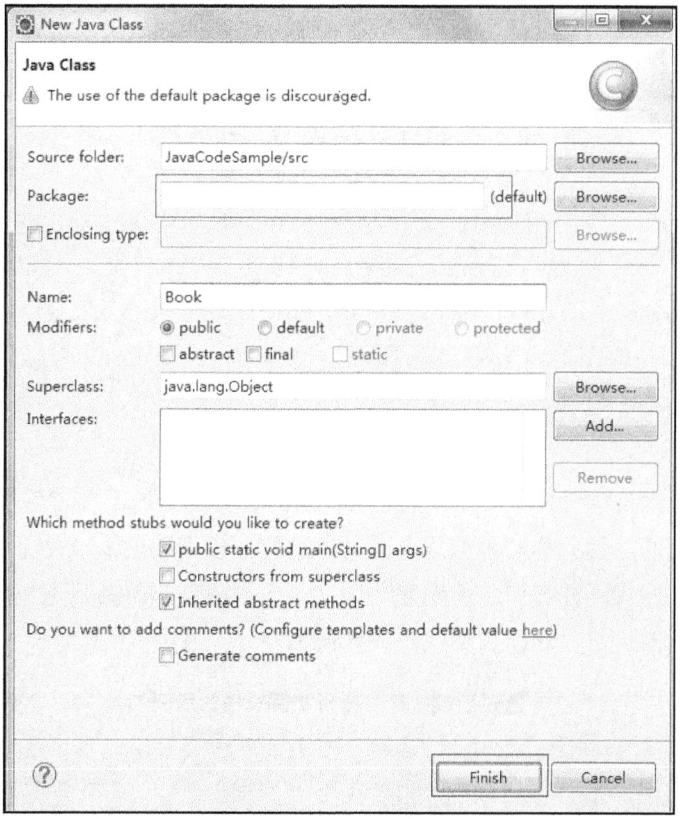

图 4-5 输入 Java 类名

其他选项皆使用默认值，单击 Finish 按钮完成类的创建。代码清单 4-1 展示了刚创建的类的代码结构。

代码清单 4-1　　　　　　　　　　新建类的示例
```java
public class Book {

    /**
     * @param args
     */
    public static void main(String[] args) {
        // TODO Auto-generated method stub

    }

}
```

人们经常进行网购,比如我们要买一本书,这本书在当当网的折扣为七折,折后的价格可以通过代码清单 4-2 计算出来。

代码清单 4-2　　　　　　　　计算书籍折扣后价格的示例
```java
    public static void main(String[] args) {
        // TODO Auto-generated method stub
        double bookPrice=30;
        double discountPrice = bookPrice * 0.7;
        System.out.println(discountPrice);
    }
```

当前代码对于定价为 30 元的书计算打七折后的价格,上述图书定价 30 元,折扣率 0.7 在代码中都是不变的,因此称这种编码方式为硬编码。硬编码的缺点是不够灵活,每次修改价格时都要修改程序代码并需要重新编译。因此,可以将图书价格提取出来,以计算对任意价格的书打七折后的价格。修改后的代码如代码清单 4-3 所示。

代码清单 4-3　　　　　　通过函数计算书籍折扣后价格的示例
```java
    /**
     * 获取折扣价格
     * @param: price 原始价格
     * @return:discountPrice 折扣后的价格
     */
```

```java
    public static double get_discount_price(double bookPrice){
        double discountPrice = bookPrice * 0.7;
        System.out.println(discountPrice);
        return discountPrice;

    }

    public static void main(String[] args) {
        get_discount_price(30);

    }
```

这个例子中的折扣率 0.7 仍然是固定的。如果要对任意价格的书按任意折扣计算优惠后的价格，目前这个函数还无法实现，因此就需要进一步将折扣率这个参数提取出来，再进行封装优化，修改后的代码如代码清单 4-4 所示。

代码清单 4-4   提取出折扣率后硬编码的示例

```java
/**
 * 获取折扣价格
 * @param: price 原始价格
 * @return:discountPrice 折扣后的价格
 */

public static double get_discount_price(double bookPrice){
    double discountPrice = bookPrice * get_discount();
    System.out.println(discountPrice);
    return discountPrice;

}

public static void main(String[] args) {
    get_discount_price(30);

}
```

在代码清单 4-4 中，将获取折扣率的方式封装成 get_discount()函数，get_discount()函数通过读取配置文件或者其他方式获得折扣率。每次折扣率发生改变时，只需要修改配置文件就可以计算书籍折扣后的价格，并不用修改程序以及重新编译。

通过上面的例子，我们可以看出在 Java 中为了提高代码的复用性，可以将函数定义成一个单独的包。函数的定义格式如下。

```
修饰符 返回值类型 函数名(参数类型 形参1,参数类型 形参1,…){
    执行语句；
    return 返回值；
}
```

当函数没有具体的返回值时，返回值的类型用 void 关键字表示。return 的作用是返回返回值，结束函数。

4.4 类

继续上面的例子，图书的折扣不可能是全场五折，有的可能是三折，有的可能不打折。折扣高低根据书的种类有所不同，如技术类书籍打五折，社科类书籍打七折。可以把有关书籍所有折扣信息的函数都放入这个 Book 类中，如代码清单 4-5 所示。

代码清单 4-5　　　　　　　　　　创建 Book 类的示例

```java
public class Book {

    /**
     * 获取技术类书籍折扣价格
     * @param: technicalBookPrice 技术类书籍原始价格
     * @return:technicalBookDiscountPrice 技术类书籍折扣后的价格
     */
    public static double get_technical_books_discount_price(double technicalBookPrice){
        double technicalBookDiscountPrice = technicalBookPrice * get_technical_books_
            discount();
        System.out.println(technicalBookDiscountPrice);
        return technicalBookDiscountPrice;
    }
}
```

```java
/**
 * 获取社科类书籍折扣价格
 * @param: socialBookPrice 社科类书籍原始价格
 * @return:socialBookDiscountPrice 社科类书籍折扣后的价格
 */
public static double get_social_books_discount_price(double socialBookPrice){
    double socialBookDiscountPrice = socialBookPrice * get_social_books_discount();
    System.out.println(socialBookDiscountPrice);
    return socialBookDiscountPrice;
}


}
```

在代码清单 4-5 中，所有书籍类型的折扣信息函数都统一放置在一个共同的 Book 类中，这样顾客在购买书的时候只要访问 Book 类找到对应的函数进行调用就可以了。例如，某顾客要购买《代码大全》，这本书属于技术类书籍，在获取书籍折扣后价格的时候调用 get_technical_books_discount_price() 函数。再比如，另一个顾客购买《我的第一本哲学书》，这本书属于社科类书籍，在获取书籍折扣后的价格的时候调用 get_social_books_discount_price()函数。

其中，get_technical_books_discount_price()和 get_social_books_discount_price()两个函数的功能和代码清单 4-3 中的 get_discount_price()一样，功能是获得折扣信息，在 4.6.1 节给出了具体的实现方法。当然，为了保持类的独立性，要把 main 函数放置到另外的类中去执行。

新建类 Customer，如代码清单 4-6 所示，假如《代码大全》的原始价格为 128 元，《我的第一本哲学书》的原始价格为 13.9 元。通过不同的函数调用获取不同的折扣率，同时返回折扣后的价格。

代码清单 4-6　　　　　　　　　　Customer 类的示例

```java
public class Customer {

    /**
     * @param args
     */
```

```java
public static void main(String[] args) {
    Book book = new Book();
    //获取《代码大全》折扣后的价格
    book.get_technical_books_discount_price(128);
    //获取《我的第一本哲学书》折扣后的价格
    book.get_social_books_discount_price(13.9);
}
```

在代码清单 4-6 中，为了调用 Book 类中的方法，首先创建了 book 对象，new 是创建对象常见的方式。

为了更好地认识世界，人们将现实生活中的事物（对象）划分成类，同一类中的事物总是具有一些共性，类以共同的特性和行为定义实体，类是具有相同属性和行为的一组对象的集合。上例的 Book 类中，书店中的书属于同一类事物，具有很大的共性，可以作为一个集合。在 Java 中，类是一个模板，它描述一类对象的行为和状态，如书是一个模板。而对象是类的一个实例，有状态和行为。例如，《代码大全》《我的第一本哲学书》是具体的书的实例，可作为一个对象。

4.5 包

类是 Java 代码封装的最高层次，为了避免出现重名的问题，Java 引入了包进行区分。例如我们网购的时候经常货比三家，看看同一本书在当当网和京东网上面的价格各为多少，从而选择更便宜的价格进行购买。而当当网和京东网都会对不同的书籍进行折扣率的配置，这里为了避免混淆就要引入包的概念。

首先，创建包。在 src 目录上右击，从上下文菜单中选择 New→Package，如图 4-6 所示。

图 4-6 新建 Java 包

4.5 包

在弹出的 New Java Package 窗口中输入包名 jd，如图 4-7 所示。

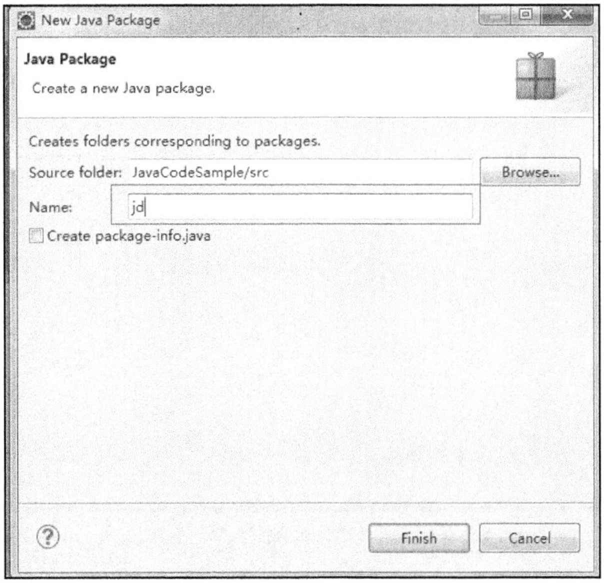

图 4-7 输入 Java 包名

使用同样的方式创建包 dangdang。将 Book 类分别复制到 jd 和 dangdang 两个包下，目录结构如图 4-8 所示。

图 4-8 创建包后的目录结构

（1）在 dangdang 包中创建 Book 类，如代码清单 4-7 所示。

代码清单 4-7　　　　　　　　当当网中 Book 类的示例

```
package dangdang;

public class Book {
```

```java
    /**
     * 获取技术类书籍折扣价格
     * @param: technicalBookPrice 技术类书籍原始价格
     * @return:technicalBookDiscountPrice 技术类书籍折扣后的价格
     */
    public static double get_technical_books_discount_price(double technicalBookPrice){
        double technicalBookDiscountPrice = technicalBookPrice * 0.5;
        System.out.println("当当网上技术书籍价格为: "+ technicalBookDiscountPrice);
        return technicalBookDiscountPrice;
    }

    /**
     * 获取社科类书籍折扣价格
     * @param: socialBookPrice 社科类书籍原始价格
     * @return:socialBookDiscountPrice 社科类书籍折扣后的价格
     */
    public static double get_social_books_discount_price(double socialBookPrice){
        double socialBookDiscountPrice = socialBookPrice * 0.7;
        System.out.println("当当网上社科书籍价格为: "+ socialBookDiscountPrice);
        return socialBookDiscountPrice;
    }
}
```

（2）在 jd 包中创建 Book 类，如代码清单 4-8 所示。

代码清单 4-8　　　　　　　　京东网中的 Book 类

```java
package jd;

public class Book {

    /**
     * 获取技术类书籍折扣价格
     * @param: technicalBookPrice 技术类书籍原始价格
     * @return:technicalBookDiscountPrice 技术类书籍折扣后的价格
     */
```

```java
    public static double get_technical_books_discount_price(double technicalBookPrice){
        double technicalBookDiscountPrice = technicalBookPrice * 0.65;
        System.out.println("京东网上技术书籍价格为: "+ technicalBookDiscountPrice);
        return technicalBookDiscountPrice;
    }

    /**
     * 获取社科类书籍折扣价格
     * @param: socialBookPrice 社科类书籍原始价格
     * @return:socialBookDiscountPrice 社科类书籍折扣后的价格
     */
    public static double get_social_books_discount_price(double socialBookPrice){
        double socialBookDiscountPrice = socialBookPrice * 0.65;
        System.out.println("京东网上社科书籍价格为: "+ socialBookDiscountPrice);
        return socialBookDiscountPrice;
    }
}
```

（3）在 Customer 类中对京东网和当当网上的书进行调用，如代码清单 4-9 所示。在调用时使用的是包名+类名的方式，从而能够避免混淆。

```java
dangdang.Book ddBook = new dangdang.Book();
jd.Book jdBook = new jd.Book();
```

代码清单 4-9　　　　　　在 Customer 类中调用 Book 类的示例

```java
public class Customer {

    /**
     * @param args
     */
    public static void main(String[] args) {
        // TODO Auto-generated method stub
        dangdang.Book ddBook = new dangdang.Book();
        //获取《代码大全》当当网折扣后的价格
```

```
        ddBook.get_technical_books_discount_price(128);
        //获取《我的第一本哲学书》当当网折扣后的价格
        ddBook.get_social_books_discount_price(13.9);
        jd.Book jdBook = new jd.Book();
        //获取《代码大全》京东网折扣后的价格
        jdBook.get_technical_books_discount_price(128);
        //获取《我的第一本哲学书》京东网折扣后的价格
        jdBook.get_social_books_discount_price(13.9);

    }

}
```

通过上面的例子可以看出，包是 Java 中为了解决类的命名问题并实现类文件的管理而引入的一种管理机制。Java 中允许将一组功能相同的类放在同一个包下，从而在逻辑上形成类的集合单元。Package 语句必须作为 Java 文件的第一条非注释语句，放在 Java 文件的第一行。一个 Java 文件只能指定到一个包下，但该文件中可以有多个类。同一个包中的类可以直接访问。

当要在一个程序里使用一个包中的类时，要用 import 子句导入包，再使用其中的类，具体语法如下。

语法 1 如下所示。

```
package ...;
import 包名.类名;
```

语法 2 如下所示。

```
package ...;
import 父包名.子包名.类名;
```

4.6 语句

方法和函数之间肯定还有很多逻辑关系，有的需要有选择地执行，有的需要重复执行，这就需要用到程序逻辑结构。下面就进行程序结构的学习。

程序结构在任何一本编程书籍里都会讲到，程序结构有三种，分别是顺序结构、分支结构和循环结构。其实在实际编程过程中能够自主编写的就只有两种，分别是分支结构和循环结构。顺序结构是作为一种程序定义而存在的，在整体上，任何一个程序，都是顺序的。无论在局部上怎么写这个程序，在整体上这个程序都是由第一个函数执行到最后一个函数。

4.6.1 条件判断

实现条件判断分支结构有两种语法，一个是 if...else 结构，一个是 switch...case 结构。

if...else 条件判断是编程语言中的语句之一。在同一个 if 结构中可以有多个 else 语句。如果第一个表达式的值为 TRUE，则执行第一个表达式；否则，执行 else 后面的内容。在上面的示例中，get_discount()函数用于获取折扣率，本节就实现这个方法。假如技术类书籍的折扣率是以书的价格为标准的，200 元以上的书籍打五折，100~200 元之间的书籍打五折，100 元以内的书籍打七折。下面通过代码清单 4-10 来实现这个函数。

代码清单 4-10   通过 if...else 实现条件判断的示例

```java
/**
 * 获取技术类书籍的折扣率
 * @param: technicalBookPrice 技术类书籍原始价格
 * @return:technicalBookDiscount 技术类书籍的折扣率
 */
public static double get_technical_book_discount(double technicalBookPrice){
    double technicalBookDiscount;
    if(technicalBookPrice >= 200){
        technicalBookDiscount = 0.5;
        System.out.println("当当网上200元以上的技术书籍折扣率为:"+technicalBookDiscount);
    }
    else if(100 <= technicalBookPrice && technicalBookPrice < 200){
        technicalBookDiscount = 0.6;
        System.out.println("当当网上100~200元之间的技术书籍折扣率为: "+technicalBookDiscount);
    }
    else{
        technicalBookDiscount = 0.7;
        System.out.println("当当网上100元以内的技术书籍折扣率为: "+technicalBookDiscount);
```

```
        }
        return technicalBookDiscount;
    }
```

当判断固定数值时,一般建议使用 switch。用小括号中变量的值依次和 case 后面的值进行对比,和哪个 case 后面的值相同就执行哪个 case 后面的语句。如果没有相同的值,则执行 default 后面的语句。假如当当网上社科类书籍的折扣率是以书的类型为标准的,A 类书籍打七折,B 类书籍打六折,其他类型的书籍打五折。下面通过代码清单 4-11 来实现这个函数。

代码清单 4-11　　　　　通过 switch…case 实现条件判断的示例

```
/**
 * 获取技术类书籍的折扣率
 * @param:bookstyle 社科类书籍类型
 * @return:socialBookDiscount 社科类书籍的折扣率
 */
public static double get_social_book_discount(char bookstyle){
    double socialBookDiscount ;
    switch(bookstyle){
    case 'A':
        socialBookDiscount = 0.7;
        System.out.println("当当网上A类社科类书籍的折扣率为: "+ socialBookDiscount);
        break;
    case 'B':
        socialBookDiscount = 0.6;
        System.out.println("当当网上B类社科类书籍折扣率为: "+ socialBookDiscount);
        break;
    default:
        socialBookDiscount = 0.5;
        System.out.println("当当网上除了A/B类社科类书籍以的外折扣率为: "+socialBookDiscount);
        break;
    }
    return socialBookDiscount;
}
```

这里，注意以下两个细节。

（1）break 是可以省略的，如果省略了，就一直执行到 break 语句。

（2）switch 后面的小括号中的变量应该是 byte、char、short、int 四种类型中的一种。

4.6.2 循环判断

Java 的循环结构提供两种语法形式：for 结构和 while 结构。

for 循环是开界的，它的一般形式为 for（初始化语句;<条件表达式>;增量定义）语句。初始化语句总是一个赋值语句，用来给循环控制变量赋初值；条件表达式是一个关系表达式，它决定什么时候退出循环；增量定义控制每循环一次后变量按什么方式变化。这三部分之间用";"分开。for 循环示例如代码清单 4-12 所示。

代码清单 4-12　　　　　　　　for 循环示例

```
//建立一个数组
int[] integers = {1, 2, 3, 4};
// 开始遍历
for (int j = 0; j<integers.length; j++){
int i = integers[j];
System.out.println(i);
}
```

while 也是一种实现循环的方式。while 循环的结构一般如下。

```
while(某种条件成立)
    {
        执行某些操作;
        改变 while 里面的那个条件;
    }
```

同样是遍历数组，采用 while 循环的方式，如代码清单 4-13 所示。

代码清单 4-13　　　　　　　　while 循环示例

```
//建立一个数组
int[] integers = {1, 2, 3, 4};
    int j = 0;
```

```
//开始遍历
        while(j < integers.length){
            int i = integers[j];
            System.out.println(i);
            j++;
        }
```

其实对于自动化测试的 Java 编程，掌握了包、类、方法和程序结构这些知识，我们就可以上手进行自动化工程的学习和实践了。当然，Java 编程中还有很多基础性内容和高级内容，希望读者再借助相关专门的书籍进行系统的学习。

4.7　Java 调试技巧

由于 Eclipse 中调试 Android 和调试 Java 的方法一样，因此这里通过调试 Java 程序来说明 Eclipse 调试技巧。

添加断点的方法比较多，一般在代码左侧边栏双击来添加断点，也可以直接按 <Ctrl+Shift+B> 快捷键，或者在左侧边栏右击并从上下文菜单中选择 Toggle BreakPoint 来创建断点，如图 4-9 所示。

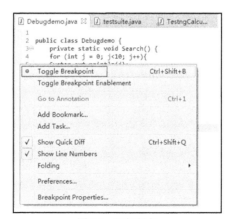

图 4-9　添加断点

切换到 Debug 模式[①]，可以在 BreakPoints 标签中查页看所有的断点列表。这里可以看到刚刚设置的断点，如图 4-10 所示。

---

① 单击 Eclipse 界面右上角的 Debug 按钮切换到 Debug 模式，单击旁边的 Java 返回普通开发模式。

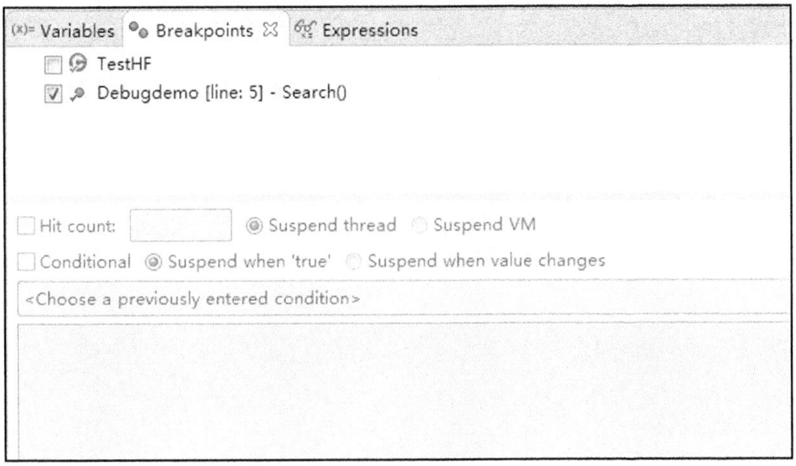

图 4-10 查看断点

常用的断点调试快捷键见表 4-1。

表 4-1 断点调试快捷键

按钮名称	快捷键	说明
Step Into	F5	单步跟踪调试程序，遇到子函数就进入并且继续单步执行
Step Over	F6	在单步执行时，在函数内遇到子函数时不会进入子函数内单步执行，而是将整个子函数执行完再停止，也就是把整个子函数作为一步
Step Return	F7	当单步执行到子函数内时，用 Step Return 就可以执行完子函数余下的部分，并返回上一层函数
Resume	F8	恢复调试，一直运行，直到遇到断点
Terminate	Ctrl+F2	中止程序的执行，关闭程序

在调试的过程中，错误提示会显示在控制台中，然后再根据提示在代码中寻找错误。

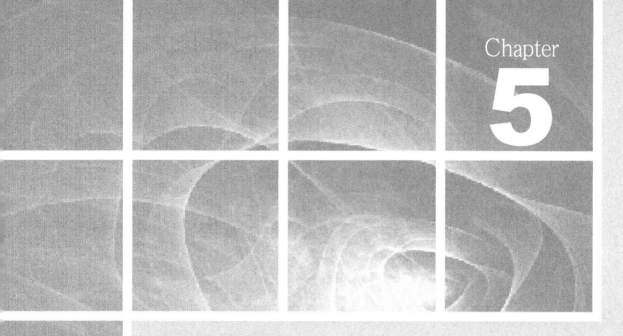

第 5 章

Android 自动化环境精讲

Android SDK 是 Google 提供的 Android 开发工具，在做 Android 测试的时候需要引入其工具包来使用相关 API。本章主要介绍 Android SDK 的安装。

Maven 是一个项目管理工具，其本身并不是一个单元测试框架，通过它可以完成 Jar 包的管理和 TestNG 测试用例的执行。

TestNG 是 Cedric Beust 创建的一种单元测试框架，程序员可以通过注解、分组等方式组织和执行自动化测试脚本。

为了全面保存日志，从多个角度展示日志信息，帮助测试人员快速定位问题，本书采用 Log4j 输出日志。

本章的目的是希望通过学习，帮助用户搭建自动化测试框架。

5.1 安装 Android SDK

Android SDK 可以从 Android Developers 官网下载。

1. 安装 Android SDK

对于 64 位的 Windows 7 操作系统，选择下载 installer_r24.4.1-windows.exe 并安装，部分安装界面如图 5-1 所示。

图 5-1　安装 Android SDK

选择 SDK 安装路径，如图 5-2 所示，进行 SDK 的安装。

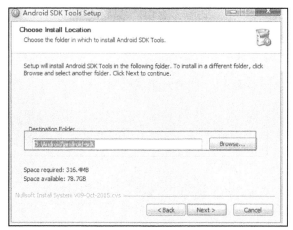

图 5-2　Android SDK 安装路径

 SDK 所在路径不能有空格。否则，Appium 提示错误"error: Logcat capture failed: spawn ENOENT"。

单击 Finish 按钮完成 Android SDK 的安装，如图 5-3 所示。

图 5-3　Android SDK 安装完成

在弹出的 Android SDK Manager 窗口中，选择需要安装的 Android API 版本以及相关的工具包。然后，单击 Install XX package 按钮（XX 表示所选的包数），如图 5-4 所示。如果要增强所做框架的系统适配性，那么建议选择所有版本的系统；如果只对某版本的系统做测试开发，则只选择对应版本的系统即可。

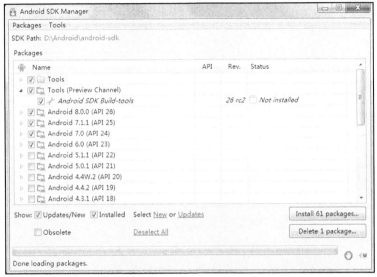

图 5-4　选择 Android API 版本

在弹出的 Choose Packages to Install 窗口中选中 Accept License 单选按钮后，单击 Install 按钮安装 API 包，如图 5-5 所示。

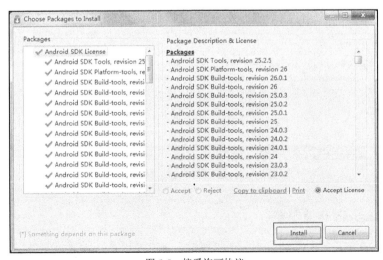

图 5-5　接受许可协议

安装过程中会显示安装日志和安装进度，如图 5-6 所示。安装时间的长短与所选包数量以及网速有关，请按照提示，耐心等待。

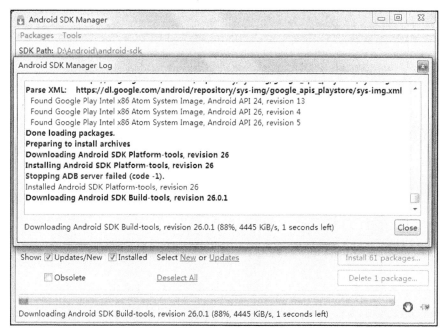

图 5-6　安装日志

当显示图 5-7 所示信息时，表示 Android SDK 安装完成。

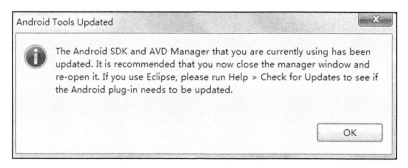

图 5-7　Android SDK 安装完成的提示

2．配置 Android 环境变量

环境变量的操作路径为"计算机"→"属性"→"高级系统设置"→"环境变量"→"系统变量"，与前边配置 JDK 环境变量的步骤一样，请参考 3.1.3 节。

（1）配置 ANDROID_HOME 环境变量。在"环境变量"对话框中，在"系统变量"选项组中，单击"新建"按钮，弹出图 5-8 所示"新建系统变量"对话框，在"变量名（N）"文本框中输入"ANDROID_HOME"，在"变量值（V）"文本框中输入 Android SDK 的安装目录。

这里 Android SDK 的安装目录为"D:\Android\android-sdk"，因此将此路径输入到"变量值（V）"文本框内。配置结果如图 5-8 所示。

图 5-8 Android SDK 环境变量的配置

（2）配置 Path 路径。选中"Path"系统环境变量，然后单击"编辑"按钮，在变量值后面添加"%ANDROID_HOME%\tools"和"%ANDROID_HOME%\platform-tools;"将 JDK 安装目录下的 bin 目录配置到 Path 环境变量中。由于 JDK 的安装目录已经配置到了"ANDROID_HOME"下，因此这里可以使用"% ANDROID_HOME %"的形式引用它，避免了二次书写。

Android SDK 下的 tools 和 platform-tools 两个目录也要配置，如图 5-9 所示。

tools 目录是 Android 开发所用到的工具，常用的有 uiautomatorviewer.bat 等。

platform-tools 目录保存一些通用工具，如 adb、aapt 等，加入环境变量有助于调用。

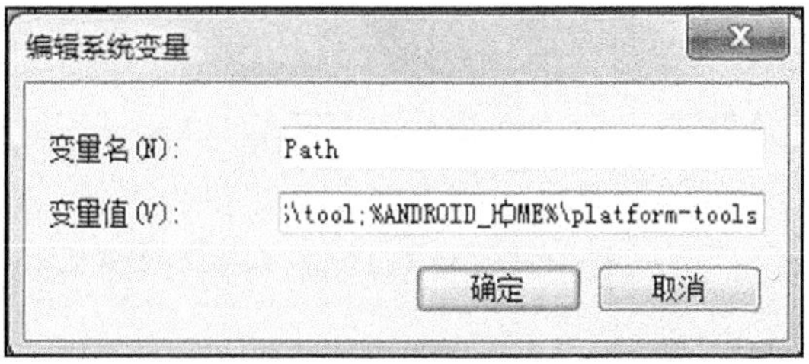

图 5-9 配置 tools 和 platform-tools 目录

5.2 Maven 项目管理

Maven 这个单词来自于意大利语，意为知识的积累，是 Apache 公司推出的一个开源项目。Maven 项目是一种方便地发布项目信息的方式，也是一种在多个项目中共享 Jar 包的方式。

使用 Java 语言编写 Appium 自动化测试框架，涉及较多的是 Jar 包的管理，以及测试框架的运行。这些操作需要通过 Maven 来完成，所以接下来介绍 Maven 的使用。

5.2.1 安装 Maven

1. 下载 Maven

安装 Maven 前要先安装 JDK 并配置 JAVA_HOME 环境变量。JDK 版本要求 1.4 及以上。关于 JDK 的具体内容参见 3.1 节。Maven 下载地址见 Maven 官网。2017 年第三季度，Maven 的最新版本为 3.5.0，如图 5-10 所示。

	Link
Binary tar.gz archive	apache-maven-3.5.0-bin.tar.gz
Binary zip archive	apache-maven-3.5.0-bin.zip
Source tar.gz archive	apache-maven-3.5.0-src.tar.gz
Source zip archive	apache-maven-3.5.0-src.zip

图 5-10 下载 Maven

根据操作系统以及个人需要，可以选择性进行下载。Maven 一般都有免安装包，所以只需要直接解压即可。如果使用的计算机采用 Windows 7 操作系统且没有其他硬性要求，可以选择 apache-maven-3.5.0-bin.zip，下载后直接解压到任意目录下即可。解压后的 Maven 项目的目录结构如图 5-11 所示。

2. 设置环境变量

环境变量的设置步骤和方法参见 3.1.3 节。具体参数及内容见表 5-1。

图 5-11 解压后 Maven 项目的目录结构

表 5-1 Maven 环境变量的配置

变　　量	值
M2_HOME	D:\09 工具\apache-maven-3.5.0-bin\apache-maven-3.5.0
M2	%M2_HOME%\bin

在"环境变量"对话框中,在"系统变量"选项组下,在 Path 系统变量中,添加";%M2%;"到字符串尾部。

 表 5-1 中 M2_HOME 的值在实际的配置中根据 Mavan 解压位置可自行修改。

3.验证环境是否安装成功

运行 cmd 命令,打开"命令提示符"界面,并输入"mvn –v"。如果 Maven 安装成功,则出现 Maven 信息,如图 5-12 所示。

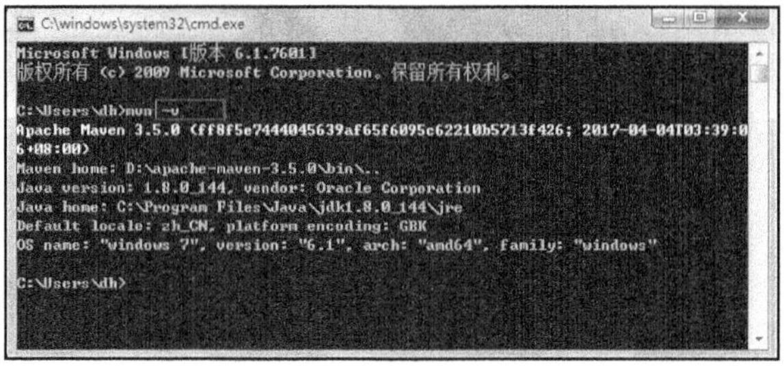

图 5-12 验证 Maven 是否安装成功

mvn 是 Maven 的一个指令，"mvn –v" 用于查看版本信息，笔者的操作系统是 64 位的 Windows 7，安装的 Maven 版本是 3.5。如果能显示如上信息，则说明到此 Maven 已经在计算机上安装完成。在 Windows 命令行中输入 "mvn help:system maven" 并按<Enter>键执行，计算机将自动从远程 Maven 库中下载大量包。

5.2.2　安装 Maven 插件

3.2.1 节介绍了 Eclipse 的安装。Maven 插件的安装是在 Eclipse 上进行的，打开 Eclipse，选择 Help→Install New Software 进行安装。

在弹出的 Install 窗口中，单击 Add 按钮。在弹出的 Add Repository 对话框中，按照图 5-13 中 Location 的设置选择插件位置。然后等待系统自动安装完成。

图 5-13　Maven 插件的安装

打开 Eclipse（若 Eclipse 已经启动则重启），在菜单栏中选择 Window→Preferences。在弹出的 Preferences 窗口中，若 Maven 插件出现在左侧面板中（见图 5-14），说明插件安装成功。

图 5-14　Maven 插件安装成功

5.2.3　创建 Maven 项目

在 Eclipse 中，在菜单栏中选择 File→New→Other。在弹出的 New 窗口中，选择 Maven→Maven Project，如图 5-15 所示。

图 5-15　创建 Maven 项目

单击 Next 按钮，弹出 New Maven Project 窗口，勾选 Use default Workspace location 复选框，如图 5-16 所示。

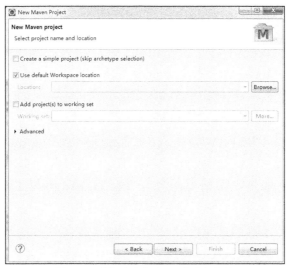

图 5-16　选择默认的项目工作空间位置

单击 Next 按钮，在图 5-17 所示的窗口中，单击 Add Archetype 按钮。

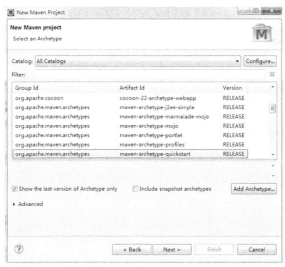

图 5-17　选择 Archetype[①]

---

① Archetype：Maven 项目的模板工具包。一个 Archetype 定义了要做的相同类型事情的初始样式或模型。

单击 Next 按钮，弹出图 5-18 所示窗口。在该窗口中，设置以下选项。

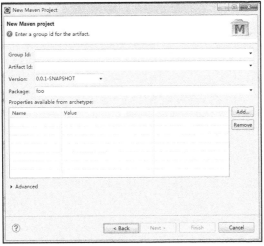

图 5-18　设置 Maven 项目的配置项

- Group Id：组织机构 ID，比如，这里填写的是"com.shijie"。如果项目不以 Jar 包的形式发布，可以自己灵活取名字。
- Artifact Id：项目打包之后 Jar 包的 ID 或者 Jar 包的名字，一般以项目名命名。
- Version：项目的版本号。在 Artifact 的仓库中，它用来区分不同的版本。

创建 Maven 项目 AppTest，填写配置项，如图 5-19 所示。

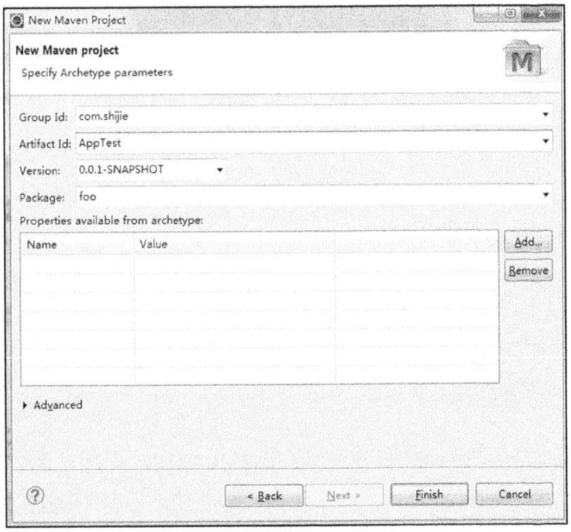

图 5-19　配置项填写完成

其他选项保持默认值即可，然后单击 Finish 按钮，项目创建成功。AppTest 项目的目录结构如图 5-20 所示。

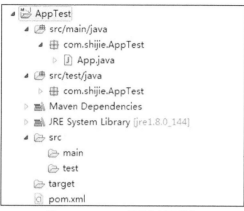

图 5-20　AppTest 项目的目录结构

Maven 提倡使用一个共同的标准目录结构，这样项目目录清晰明了，见表 5-2。

表 5-2　Maven 项目的目录结构

目录结构	描述
src/main/java	项目的源代码所在的目录
src/test/java	测试代码所在的目录
LICENSE.txt	项目的许可文件
README.txt	项目的 readme 文件
target	存放项目构建后的文件和目录以及 Jar 包、War 包、编译的 Class 文件等
pom.xml	项目的描述文件

5.2.4　Maven 项目依赖包

Maven 是基于项目对象模型（POM）并且可以通过一小段描述信息来管理项目的构建、报告和文档的软件项目管理工具。

Maven 项目创建成功后，项目目录下会有个 pom.xml 文件，通过 pom.xml 文件，添加项目所必需的 Jar 包。用户不需要刻意去下载 Jar 包，只需要一段代码即可，其他工作交给 Maven 去处理（Maven 会根据 Jar 包依赖从镜像中心查找相关 Jar 包）。

5.2.5　Maven 坐标定位

为了使 Maven 自动下载 Jar 包，首先要明白 Maven 的坐标定位。Maven 的版本控制是使用 Maven 坐标来完成的，而 Maven 坐标主要由 GAV（Groupid、ArtifactId、Version）构成。因此，使用任何一个依赖之前，用户都需要知道它的 Maven 坐标。最简单的 Maven 坐标依赖如代码清单 5-1 所示。

代码清单 5-1　　　　　　　　Maven 坐标依赖

```
<dependency>
    <groupId>junit</groupId>
    <artifactId>junit</artifactId>
    <version>4.12</version>
</dependency>
```

代码清单 5-1 中涉及几个名词，表 5-3 对其进行了详细的解释。

表 5-3　Maven 坐标详解

坐标	描述
groupId	顾名思义，这个应该是公司名或者组织名。一般来说，groupId 由三部分组成，每个部分之间以"."分隔。第一部分是项目用途，比如用于商业的就是 com，用于非营利性组织的就是 org；第二部分是公司名，比如 tengxun、baidu、alibaba；第三部分是项目名
artifactId	Maven 构建的项目名，当项目中有子项目时，就可以使用"项目名-子项目名"的命名方式
version	版本号，该元素定义 Maven 项目当前所处的版本，如代码清单 5-1 中 junit 的版本为 4.1.2。如果该值为 SNAPSHOT，说明该项目还在开发中，是不稳定的版本

如果向 pom.xml 文件中添加一个 dependency 的坐标，并使用<Ctrl+S>快捷键保存，Maven 就会自动把需要的 Jar 包和对应的依赖包下载和配置好。下载的位置默认为用户的目录下名为/.m2/repository/的仓库目录。

单击进入 pom.xml 文件，切换至 pom.xml 选项卡，然后把代码清单 5-1 粘贴在<dependencies></dependencies>之间，单击"保存"按钮，Maven 会自动下载依赖包，如图 5-21 所示。

图 5-21　通过 Maven 自动下载依赖包

保存之后，我们发现 Maven 自动开始下载对应版本的 Jar 包。当下载完之后，在 Maven 项目的目录结构中，单击 Maven Dependencies 左侧的小三角形，发现里面有相关 Jar 包了。

有些人肯定会问：如何知道各个 Jar 包的依赖代码，这里有个简单的方法，Maven 会有各个软件服务商所提供的 Maven 仓库，在这个仓库中查询即可。当输入 TestNG 时，Maven 仓库会返回 TestNG 所有版本的依赖代码。

这里提供一个 Maven 仓库（互联网中有很多 Maven 仓库，用户可以根据需要自由选择）。在如图 5-22 所示界面中，搜索"TestNG"，选择所需要的版本，将给出的 Maven 坐标复制到 pom.xml 中即可使用 Maven 仓库。

图 5-22　常用 Maven 仓库

5.3　TestNG 测试框架简介

使用 Java 语言编写的 Appium 自动化测试框架通常是使用单元测试框架运行的，所以读者有必要了解单元测试框架的基本使用方法和单元测试框架的使用技巧。

5.3.1　安装 TestNG

TestNG 是 Java 中的一个单元测试框架，类似于 JUnit[①]，是一种注解式的编程方式，支

---

① JUnit：基于 Java 语言的主流单元测试框架。

持并行运行、数据驱动等。本章介绍常用的 TestNG 方法，用于满足自动化测试的常用要求。具体内容请参考详细的 TestNG 说明文档。

安装 TestNG 的步骤如下所示。

（1）打开 Eclipse IDE，在菜单栏中选择 Help→Eclipse Marketplace，弹出 Eclipse Marketplace 窗口。在该窗口中，在 Find 文本框中输入 TestNG，如图 5-23 所示。

图 5-23　查找 TestNG 软件

（2）找到 TestNG 软件后，单击对应的 Install 按钮，开始下载安装，如图 5-24 所示。

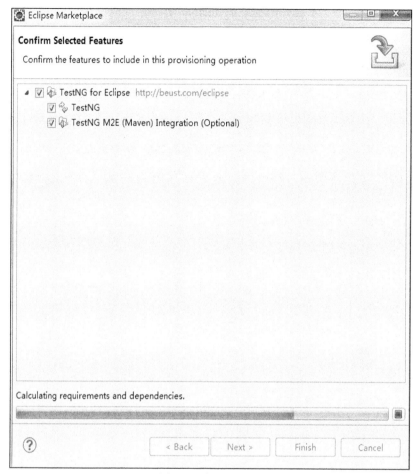

图 5-24 安装 TestNG

安装成功后，根据提示，选择是否重新启动 Eclipse，如图 5-25 所示。

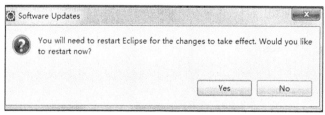

图 5-25 TestNG 插件安装完成

Eclipse 重启后，右击任意项目名，如果在弹出的上下文菜单中有 TestNG 选项，表示 TestNG 安装成功，如图 5-26 所示。

图 5-26　验证 TestNG 安装完成

5.3.2　TestNG 测试用例

在 5.2.3 节中，创建了 AppTest 这个 Maven 项目。本节基于此项目创建 TestNG 测试类，AppTest 项目的目录结构如图 5-27 所示。

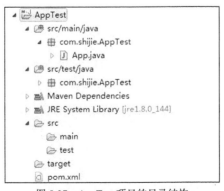

图 5-27　AppTest 项目的目录结构

1. 创建待测类

在创建测试用例之前，先创建待测程序。打开已经创建的 Maven 项目 AppTest，在 src/main/java 目录下，右击包 com.shijie.AppTest，选择 New→Class 菜单项。在 Name 文本框中输入 Calculate，然后单击"Finish"按钮创建 Java 类，如图 5-28 所示。

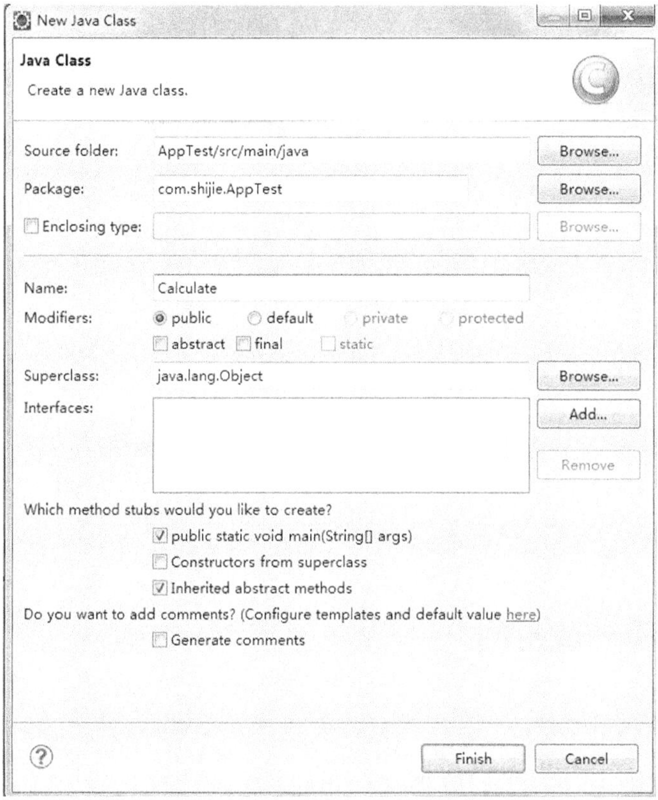

图 5-28 创建待测类

在 Calculate 类中输入代码，如代码清单 5-2 所示。

代码清单 5-2　　　　　　　　　　待测类

```
package com.shijie.AppTest;

public class Calculate {

    /**
     * @param num1/num2 两个加数
```

```
 * @return 加数之和
 */
public int add(int num1, int num2){
    return num1 + num2;
}
}
```

这是一个非常简单的 Java 函数，实现加法操作，传入两个参数，返回两个参数的和。

2. 配置 TestNG 测试环境

在 AppTest 项目的目录结构中，右击 AppTest，从上下文菜单中选择 Build Path→Configure Build Path。在弹出的 Properties for AppTest 对话框中，在左侧面板中，选择 Java Build Path，在右侧面板中，选择 Libraries 选项卡，单击 Add Library 按钮，如图 5-29 所示。

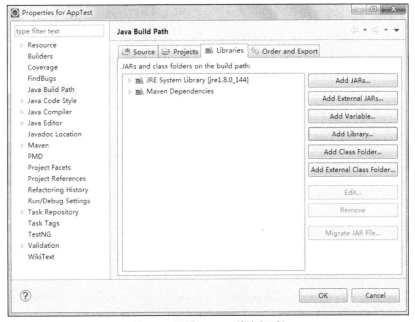

图 5-29　配置 TestNG 到测试工程

在弹出的 Add Library 窗口中选择 TestNG，单击 Next 按钮。回到 Properties for AppTest 对话框，单击 OK 按钮，TestNG 配置完成。配置成功后，AppTest 项目的目录下会出现 TestNG 目录，如图 5-30 所示。

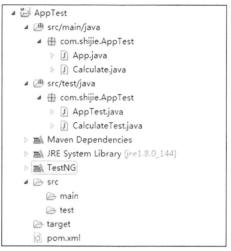

图 5-30　成功配置 TestNG 到项目中

3．创建测试类

在 src/test/java 目录下，右击包 com.shijie.AppTest，从上下文菜单中依次选择 TestNG→Create TestNG class，创建 CalculateTest 测试类，弹出图 5-31 所示对话框。

图 5-31　创建测试类

如图 5-31 所示，在 Class name 文本框中输入 CalculateTest，Annotations 选项组中为 TestNG 的注解，这里除了 @DataProvider 全部勾选，本节将依次介绍这些注解的作用。DataProvider 用于实现参数化，这会在 5.3.3 节中专门讲述。创建测试类 CalculateTest 成功后，输入测试代码，如代码清单 5-3 所示。

代码清单 5-3　　　　　　　　　　测试代码

```java
package com.shijie.AppTest;

import org.testng.Assert;
import org.testng.annotations.Test;
import org.testng.annotations.BeforeMethod;
import org.testng.annotations.AfterMethod;
import org.testng.annotations.BeforeClass;
import org.testng.annotations.AfterClass;
import org.testng.annotations.BeforeTest;
import org.testng.annotations.AfterTest;
import org.testng.annotations.BeforeSuite;
import org.testng.annotations.AfterSuite;

public class CalculateTest {
  @Test
  public void f() {
      Calculate calc = new Calculate();
      int result;
      result = calc.add(1, 2);
      Assert.assertEquals(result, 3);
  }
  @BeforeMethod
  public void beforeMethod() {
      System.out.println("BeforeMethod");
  }

  @AfterMethod
  public void afterMethod() {
      System.out.println("AfterMethod");
  }
```

```java
@BeforeClass
public void beforeClass() {
    System.out.println("BeforeClass");
}

@AfterClass
public void afterClass() {
    System.out.println("AfterClass");
}

@BeforeTest
public void beforeTest() {
    System.out.println("BeforeTest");
}

@AfterTest
public void afterTest() {
    System.out.println("AfterTest");
}

@BeforeSuite
public void beforeSuite() {
    System.out.println("BeforeSuite");
}

@AfterSuite
public void afterSuite() {
    System.out.println("AfterSuite");
}
}
```

4．运行测试脚本

右击测试类 CalculateTest，并从上下文菜单中选择 Run As→TestNG Test，执行测试用例。TestNG 运行结果如图 5-32 所示。

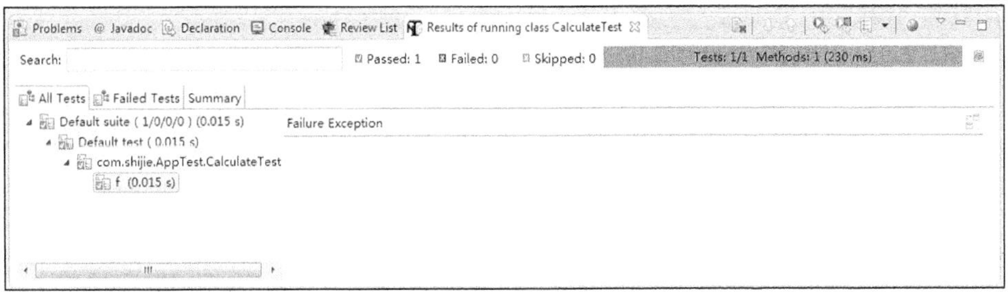

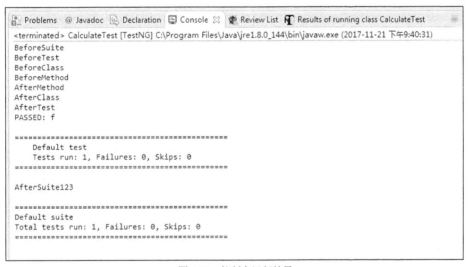

图 5-32　TestNG 运行结果

控制台运行结果如图 5-33 所示。

图 5-33　控制台运行结果

图 5-33 中可以看到这些注解的运行顺序，分辨出不同的注解会在何时调用。

在 TestNG 中常见的测试用例组织结构如下。

- 一个 TestSuite 由多个 Test 组成。
- 一个 Test 由多个 Class 组成。
- 一个 Class 由多个 Method 组成。

当运行不同层级的测试用例时，可通过不同的注解实现测试前的初始化工作、测试用例的执行工作和测试执行后的环境清理工作。测试运行前，要打开文件、启动数据库、读取参数配置。测试运行结束后，要关闭文件、关闭数据库、清理运行环境等。

常用注解见表 5-4。

表 5-4 常用注解

注　　解	说　　明
@BeforeSuite	当前测试集合（Suite）中任意测试用例开始运行之前执行
@AfterSuite	当前测试集合中所有测试用例运行结束之后执行
@BeforeTest	Test 中任意测试用例开始运行之前执行
@AfterTest	Test 中所有测试用例运行结束之后执行
@BeforeClass	当前测试类（Class）中任意测试用例开始运行之前执行
@AfterClass	当前测试类中所有测试用例运行结束之后执行
@BeforeMethod	每个测试方法（Method）开始运行之前执行
@AfterMethod	每个测试方法运行结束之后执行

5.3.3 数据驱动

自动化测试必然有测试数据，测试数据有多有少。相同的脚本使用不同的测试数据来执行，实现了测试数据和测试行为的完全分离。TestNG 有以下几种数据驱动方式。

- 直接在 testng.xml 文件中定义简单参数，然后在源文件中引用这些参数。这适用于测试数据少的情况。
- 直接将测试数据写到测试类中，然后使用 dataProvider 来关联数据。这适用于测试数据少的情况。
- 将数据写到外部数据文件（如 excel、cvs 等）中。这适用于数据量多的情况。

我们仍然以 Calculate 待测类为例，设计测试用例完成对 Add 方法的测试。测试用例的设计见表 5-5，其中设计了两个正常用例和一个异常用例。

表 5-5 测试用例设计

用 例 标 号	参　数　1	参　数　2	期 望 结 果
1	1	2	3
2	1	999	1000
3	4	2147483644	抛出溢出异常

下面分别通过不同的方法去实现这个测试用例。

1. 通过 testng.xml 传递参数

如果只使用相对简单的参数，那么用户可以在 testng.xml 文件中指定。

在这段代码中，使 num1 和 num2 参数能够分别接收到 testng.xml 文件中名为 param1 和 param2 参数传来的值。这两个参数都定义在 testng.xml 文件中。

TestNG 的 testng.xml 配置文件用来辅助定义执行什么样的测试，通过直接创建 testNG 类可能不会产生 XML 文件，需要手工配置。

下面介绍如何配置 testng.xml。

5.3.2 节中运行了 TestNG 测试用例，这会在环境变量 tmp 所表示的目录下产生一个运行临时文件夹。在"运行"命令对话框中输入"%tmp%"，打开 Temp 临时文件夹，如图 5-34 所示。

图 5-34 打开 tmp 临时文件夹

打开 Temp 文件夹后，找到以 testng-eclipse 开头的文件夹，如图 5-35 所示。

图 5-35 TestNG 临时运行文件夹

打开这个文件夹后，下面会有一个 XML 文件，如图 5-36 所示。

图 5-36 TestNG 临时运行文件

使用记事本打开并查看该文件，如代码清单 5-4 所示。

代码清单 5-4　　　　　　　　　　TestNG 临时运行文件

```xml
<?xml version="1.0" encoding="UTF-8"?>
<!DOCTYPE suite SYSTEM "http://testng/testng-1.0.dtd">
<suite name="Default suite">
  <test verbose="2" name="Default test">
    <classes>
      <class name="com.shijie.AppTest.CalculateTest"/>
    </classes>
  </test> <!-- Default test -->
</suite> <!-- Default suite -->
```

代码清单 5-4 符合 TestNG 测试用例的组织结构，其中包含 suite、test、class 等字段，具体参见 5.3.2 节。

<suite>元素是 testng.xml 文件的根元素。一个<suite>可以包含多个<test>元素。一个<suite>可以包含一个<groups>元素，用于定义全局的组，该组对所有的测试可见。

<test>元素是<suite>的子元素，用于定义一个测试用例。一个<test>中可以包括多个<class>或者多个<package>。

<class>是<test>的子元素，用于定义一个测试类。一个<class>中可以包括多个<method>。

<method>中可以指定具体待运行的测试方法，通过 include 和 exclude 选择或者排除具体的测试方法。

该 XML 文件与 TestNG 自动创建的运行 XML 文件一致，可以直接将该文件复制到项目目录下并修改和使用。修改后的 XML 文件如代码清单 5-5 所示。

代码清单 5-5　　　　　　　　　　TestNG.xml 文件

```xml
<?xml version="1.0" encoding="UTF-8"?>
<!DOCTYPE suite SYSTEM "http://testng/testng-1.0.dtd">
<suite name="Default suite">
  <test verbose="2" name="Default test">
    <parameter name="param1" value="1"/>
    <parameter name="param2" value="2"/>
    <classes>
      <class name="com.shijie.AppTest.CalculateTest"/>
    </classes>
```

```
</test> <!-- Default test -->
</suite> <!-- Default suite -->
```

在 test 中定义了两个参数 param1 和 param2。关于参数的有效范围，需要注意以下几点。

（1）在 Suite 范围内定义的某个参数的值，对所有的 Test 都有效。

（2）在 Test 范围内定义的某个参数的值，只针对该 Test 有效。

（3）如果同时在 Suite 和 Test 中定义某个参数，Test 范围内的值会屏蔽 Suite 的值。

例如，在测试中只运行方法 f()，修改测试类 CalculateTest，如代码清单 5-6 所示。使用了注解@Parameters。这个注解可以从 TestNG 外部传入参数，如在 testng.xml 文件中定义的参数。在@Parameters 注解中指定参数列表（param1 和 param2），同时测试方法 f()中要声明对应的形参（num1 和 num2），形参和参数列表要一一对应，名称可以不同。

代码清单 5-6　　　　　　　　　　测试类

```
package com.shijie.AppTest;

import org.testng.Assert;
import org.testng.annotations.Parameters;
import org.testng.annotations.Test;
import org.testng.annotations.BeforeMethod;
import org.testng.annotations.AfterMethod;
import org.testng.annotations.BeforeClass;
import org.testng.annotations.AfterClass;
import org.testng.annotations.BeforeTest;
import org.testng.annotations.AfterTest;
import org.testng.annotations.BeforeSuite;
import org.testng.annotations.AfterSuite;

public class CalculateTest {
  @Test
  @Parameters({"param1" , "param2"})
  public void f(int num1, int num2) {
      Calculate calc = new Calculate();
      int result;
      result = calc.add(num1, num2);
```

```java
        Assert.assertEquals(result, 3);
}

@BeforeMethod
public void beforeMethod() {
    System.out.println("BeforeMethod");
}

@AfterMethod
public void afterMethod() {
    System.out.println("AfterMethod");
}


@BeforeClass
public void beforeClass() {
    System.out.println("BeforeClass");
}

@AfterClass
public void afterClass() {
    System.out.println("AfterClass");
}

@BeforeTest
public void beforeTest() {
    System.out.println("BeforeTest");
}

@AfterTest
public void afterTest() {
    System.out.println("AfterTest");
}

@BeforeSuite
```

```java
    public void beforeSuite() {
        System.out.println("BeforeSuite");
    }

    @AfterSuite
    public void afterSuite() {
        System.out.println("AfterSuite");
    }
}
```

 在 XML 数据驱动方式中，运行入口为 testng.xml 文件，具体操作为右击 testng.xml 文件，从上下文菜单中选择 Run as→TestNG Suite。

2. 通过 DataProvider 传递参数

testng.xml 只用来传递简单的参数。如果需要创建复杂的参数，或者从 Java 中创建参数，可以使用 DataProvider 来给需要的测试提供参数。所谓数据提供者，就是能返回对象数组的方法，并且这个方法被@DataProvider 注解标注。

在 5.3.2 节创建 TestNG 测试类的时候未勾选@DataProvider 复选框。这里创建测试类 CalculateTest2，在创建的时候勾选@DataProvider 复选框。创建结果如代码清单 5-7 所示。

代码清单 5-7　　　　　　　　　勾选数据提供者

```java
package com.shijie.AppTest;

import org.testng.annotations.Test;
import org.testng.annotations.BeforeMethod;
import org.testng.annotations.AfterMethod;
import org.testng.annotations.DataProvider;
import org.testng.annotations.BeforeClass;
import org.testng.annotations.AfterClass;
import org.testng.annotations.BeforeTest;
import org.testng.annotations.AfterTest;
import org.testng.annotations.BeforeSuite;
import org.testng.annotations.AfterSuite;
```

```java
public class CalculateTest {
  @Test(dataProvider = "dp")
  public void f(Integer n, String s) {
  }
  @BeforeMethod
  public void beforeMethod() {
  }

  @AfterMethod
  public void afterMethod() {
  }


  @DataProvider
  public Object[][] dp() {
    return new Object[][] {
      new Object[] { 1, "a" },
      new Object[] { 2, "b" },
    };
  }
  @BeforeClass
  public void beforeClass() {
  }

  @AfterClass
  public void afterClass() {
  }

  @BeforeTest
  public void beforeTest() {
  }

  @AfterTest
```

```java
    public void afterTest() {
    }


    @BeforeSuite
    public void beforeSuite() {
    }


    @AfterSuite
    public void afterSuite() {
    }
}
```

DataProvider 注解定义返回的是测试数据集，这个测试数据集是一个关于 Object 的二维数组。其中的每个一维数组都会传递给调用函数，并作为参数使用。在运行的时候，我们会发现@Test 标识的 Test Method 被执行的次数和 Object[][]包含的一维数组的个数是一致的，而@Test 标识的函数的参数个数，也和 Object 内一维数组内的元素数是一致的。代码清单 5-7 命名为"dp"。

被@Test 标注的方法通过 DataProvider 属性指明其数据提供商。这个名字必须与@DataProvider(name="...")中的名字相一致。另外，被@Test 标注的方法要声明对应的形参（Integer n, String s），形参和参数列表要一一对应，名称可以不同。修改测试类，完成对 add 方法的测试，如代码清单 5-8 所示。

代码清单 5-8　　　　　　　　　　修改后的测试类

```java
package com.shijie.AppTest;

import org.testng.Assert;
import org.testng.annotations.Test;
import org.testng.annotations.BeforeMethod;
import org.testng.annotations.AfterMethod;
import org.testng.annotations.DataProvider;
import org.testng.annotations.BeforeClass;
import org.testng.annotations.AfterClass;
import org.testng.annotations.BeforeTest;
import org.testng.annotations.AfterTest;
import org.testng.annotations.BeforeSuite;
```

```java
import org.testng.annotations.AfterSuite;


public class CalculateTest {
  @Test(dataProvider = "dp")
    public void f(int n, int s) {
        Calculate calc = new Calculate();
        int result;
        result = calc.add(n, s);
        System.out.println(result);
    }


    @DataProvider
    public Object[][] dp() {
      return new Object[][] {
        new Object[] { 1, 2 },
        new Object[] { 2, 999 },
        new Object[] { 4, 2147483644 },
      };
    }
    @BeforeMethod
    public void beforeMethod() {
        System.out.println("BeforeMethod");
    }


    @AfterMethod
    public void afterMethod() {
        System.out.println("AfterMethod");
    }


    @BeforeClass
    public void beforeClass() {
        System.out.println("BeforeClass");
    }
```

```java
@AfterClass
public void afterClass() {
    System.out.println("AfterClass");
}

@BeforeTest
public void beforeTest() {
    System.out.println("BeforeTest");
}

@AfterTest
public void afterTest() {
    System.out.println("AfterTest");
}

@BeforeSuite
public void beforeSuite() {
    System.out.println("BeforeSuite");
}

@AfterSuite
public void afterSuite() {
    System.out.println("AfterSuite");
}
}
```

测试方法 f()中的两个参数分别使用 dp 测试数据集中每个一维数组的数据进行赋值，此方法被调用了 3 次，分别使用 dp 测试数据集中的 3 组数据。

第一次调用使用的参数值如下。

```
n: 1
s: 2
```

第二次调用使用的参数值如下。

```
n: 2
s: 999
```

第三次调用使用的参数值如下。

```
n: 4
s: 2147483644
```

根据 TestNG 运行结果可知，测试用例的执行结果都是 PASSED。通过输出的数值可以看出，第三次运行的测试用例结果为 2147483644，产生了数据溢出。待测类因为没有加入异常所以忽略了该问题的存在，同时测试类因为没有增加异常测试也忽略了这个错误。异常测试会在 5.3.9 节中详细讲述。

5.3.4 分组测试

TestNG 通过 group 关键字来完成测试方法的分组，使用 testng.xml 中的<group>字段来指定分组。本节涉及两个新的 TestNG 注解，见表 5-6。

表 5-6 新的 TestNG 注解

注解	说明
@BeforeGroup	配置方法将在之前运行组列表。此方法保证在调用其中任何一组的第一个测试方法之前运行
@AfterGroup	此配置方法将在之后运行组列表。该方法保证在调用其中任何一组的最后一个测试方法之后运行

修改 Calculate 待测类，如代码清单 5-9 所示，增加待测方法。

代码清单 5-9　　　　　　　　　　待测类

```java
package com.shijie.AppTest;

public class Calculate {

    /**
     * 加法运算
     * @param num1/num2 两个加数
     * @return 两数之和
     */
    public int add(int num1, int num2){
        return num1 + num2;
```

```java
    }

    /**
     * 减法运算
     * @param num1/num2，即被减数和减数
     * @return 两数之差
     */
    public int subtraction(int num1, int num2){
        return num1 - num2;

    }

    /**
     * 猫介绍
     */
    public void dog(){
        System.out.println("我是一只猫");

    }

    /**
     * 狗介绍
     */
    public void cat(){
        System.out.println("我是一只狗");

    }

    /**
     * 绿萝介绍
     */
    public void scindapsus(){
        System.out.println("我是一盆绿萝");
```

```
    }

    /**
     * 君子兰介绍
     */
    public void clivia(){
        System.out.println("我是一盆君子兰");
    }

}
```

同时创建测试类 CalculateTest，通过 group 关键字完成分组，如代码清单 5-10 所示。分组示例如下。

（1）addTest()方法和 subtractionTest()方法属于"运算"分组。

（2）dogTest()方法和 catTest()方法属于"动物"分组。

（3）scindapsusTest()方法和 cliviaTest()方法属于"植物"分组。

（4）dogTest()方法、catTest()方法、scindapsusTest()方法、cliviaTest()方法属于"非运算"分组。

代码清单 5-10　　　　　　　　　　测试类

```
package com.shijie.AppTest;

import org.testng.annotations.Test;

public class CalculateTest {
    @Test(groups = {"运算"})
    public void addTest() {
        System.out.println("加法测试");
    }

    @Test(groups = {"运算"})
    public void subtractionTest() {
        System.out.println("减法测试");
    }
```

```java
@Test(groups = {"动物","非运算"})
public void dogTest() {
    System.out.println("测试狗");
}
@Test(groups = {"动物","非运算"})
public void catTest() {
    System.out.println("测试猫");
}
@Test(groups = {"植物","非运算"})
public void scindapsusTest() {
    System.out.println("测试绿萝");
}
@Test(groups = {"植物","非运算"})
public void cliviaTest() {
    System.out.println("测试君子兰");
}
}
```

testng.xml 的配置内容如下。

```xml
<?xml version="1.0" encoding="UTF-8"?>
<!DOCTYPE suite SYSTEM "http://testng/testng-1.0.dtd">
<suite name="Default suite">
  <test verbose="2" name="Default test">
    <groups>
      <run>
          <include name = "非运算"/>
      </run>
    </groups>
    <classes>
      <class name="com.shijie.AppTest.CalculateTest3"/>
    </classes>
  </test> <!-- Default test -->
</suite> <!-- Default suite -->
```

右击 testng.xml 文件，从上下文菜单中选择 Run as→TestNG Suite 运行测试。运行结果如图 5-37 所示。

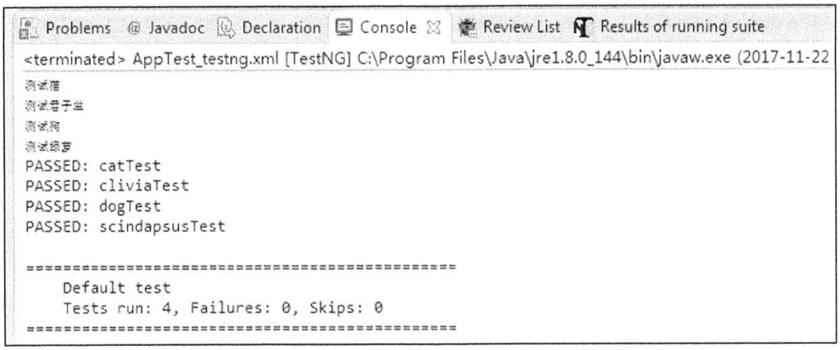

```
Problems @ Javadoc  Declaration  Console ⊠  Review List  Results of running suite
<terminated> AppTest_testng.xml [TestNG] C:\Program Files\Java\jre1.8.0_144\bin\javaw.exe (2017-11-22
测试猫
测试吊子兰
测试狗
测试绿萝
PASSED: catTest
PASSED: cliviaTest
PASSED: dogTest
PASSED: scindapsusTest

===============================================
    Default test
    Tests run: 4, Failures: 0, Skips: 0
===============================================
```

图 5-37　运行结果

从图 5-37 中运行结果可以看出，所有分组为"非运算"的方法都运行，修改 testng.xml 配置文件可以实现运行不同的分组。

5.3.5　按照特定顺序执行测试用例

TestNG 测试用例的运行结果是无序的，每次都按照不同的顺序运行。修改代码清单 5-10，通过 priority 参数修改测试用例的执行顺序。例如，先执行"动物"分组的方法（catTest、dogTest），再执行"植物"分组的方法（scindapsusTest、cliviaTest），最后执行"运算"分组的方法（addTest、subtractionTest），如代码清单 5-11 所示。

代码清单 5-11　　　　　　　　测试类

```java
package com.shijie.AppTest;

import org.testng.annotations.Test;

public class CalculateTest {
    @Test(priority = 4)
    public void addTest() {
        System.out.println("加法测试");
    }
    @Test(priority = 5)
    public void subtractionTest() {
        System.out.println("减法测试");
    }
```

```
@Test(priority = 1)
public void dogTest() {
    System.out.println("测试狗");
}
@Test(priority = 0)
public void catTest() {
    System.out.println("测试猫");
}
@Test(priority = 2)
public void scindapsusTest() {
    System.out.println("测试绿萝");
}
@Test(priority = 3)
public void cliviaTest() {
    System.out.println("测试君子兰");
}
}
```

运行结果如图 5-38 所示，按照指定的顺序运行。

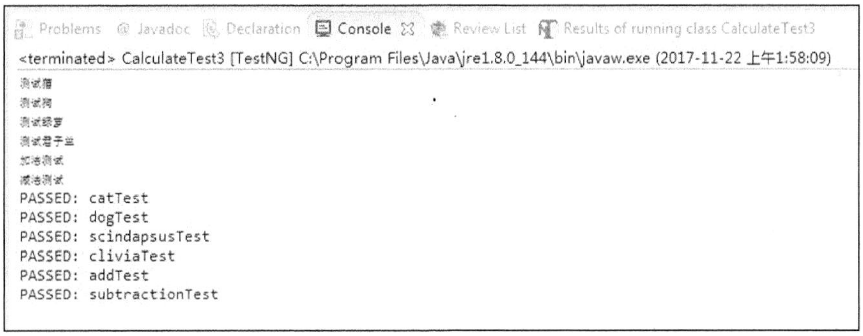

图 5-38　运行结果

> priority 值越小，运行级别就越高，如 priority 为 0 的方法比 priority 为 5 的方法运行得要早。

5.3.6 忽略测试

当测试用例没有书写完成或者不需要执行时，可以采用注释@Test(enabled = false)来禁止测试用例的执行。修改代码清单 5-11，禁止运行所有的"运算"方法（addTest 和 subtractionTest），如代码清单 5-12 所示。

代码清单 5-12　　　　　　　　　　忽略测试类

```java
package com.shijie.AppTest;

import org.testng.annotations.Test;

public class CalculateTest {
    @Test(priority = 4, enabled = false)
    public void addTest() {
        System.out.println("加法测试");
    }
    @Test(priority = 5, enabled = false)
    public void subtractionTest() {
        System.out.println("减法测试");
    }
    @Test(priority = 1)
    public void dogTest() {
        System.out.println("测试狗");
    }
    @Test(priority = 0)
    public void catTest() {
        System.out.println("测试猫");
    }
    @Test(priority = 2)
    public void scindapsusTest() {
        System.out.println("测试绿萝");
    }
    @Test(priority = 3)
```

```java
    public void cliviaTest() {
        System.out.println("测试君子兰");
    }
}
```

运行结果如图 5-39 所示,"运算"分组的 addTest 和 subtractionTest 方法没有运行。

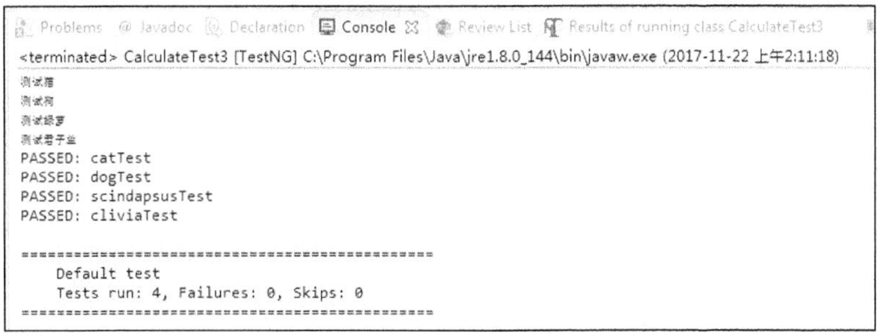

图 5-39 忽略测试运行结果

5.3.7 依赖测试

我们可能需要以特定顺序调用测试用例中的方法,或者可能希望在方法之间共享一些数据和状态。例如,要在论坛中发帖子,前提是登录成功,所以发帖子的方法依赖于登录成功方法的执行。使用 dependsOnMethods 或者 dependsOnGroups 来指定依赖关系。

修改后的测试类如代码清单 5-13 所示。

代码清单 5-13　　　　　　　　　依赖测试类

```java
package com.shijie.AppTest;

import org.testng.annotations.Test;

public class CalculateTest {

    @Test(dependsOnMethods = {"loginTest"})
    public void postTest() {
        System.out.println("发帖");
    }
}
```

```java
@Test
public void loginTest() {
    System.out.println("登录");
}
}
```

运行结果如图 5-40 所示，在发帖方法被调用之前先调用了登录方法，从而实现依赖测试。

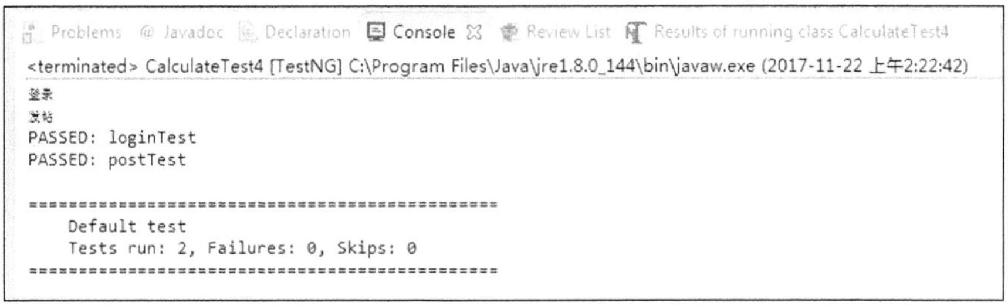

图 5-40　依赖测试运行结果

5.3.8　超时测试

"超时"表示如果单元测试花费的时间超过指定的毫秒数，那么 TestNG 将会中止它并将其标记为失败。修改后的测试类如代码清单 5-14 所示。OntimeTest 方法的测试运行时间（1000ms）短于超时时间（2000ms），timeoutTest 方法的测试运行时间（3000ms）长于超时时间（2000ms）。

代码清单 5-14　　　　　　　　　　超时测试类

```java
package com.shijie.AppTest;

import org.testng.annotations.Test;

public class CalculateTest {
    @Test(timeOut = 2000)
    public void timeoutTest() throws InterruptedException {
        Thread.sleep(3000);
    }
}
```

```
@Test(timeOut = 2000)
public void ontimeTest() throws InterruptedException {
    Thread.sleep(1000);

}
}
```

运行结果如图 5-41 所示，ontimeTest 方法运行成功，timeoutTest 方法在运行时报错。

图 5-41　超时测试运行结果

5.3.9　异常测试

在测试 add 方法时，测试用例的设计包括 2147483647 加上 4，因为整型数的最大值为 2147483647，所以 2147483647+4 已经超过了整型数的范围。当编写正规的代码时，应该抛出溢出错误，防止多层次的计算导致的数据误差。TestNG 采用@Test 注释 expectedExceptions 参数来验证异常是否抛出。修改后的测试类如代码清单 5-15 所示。

代码清单 5-15　　　　　　　　　异常测试类

```
package com.shijie.AppTest;

import org.testng.Assert;
import org.testng.annotations.Test;
import org.testng.annotations.BeforeMethod;
import org.testng.annotations.AfterMethod;
import org.testng.annotations.DataProvider;
import org.testng.annotations.BeforeClass;
```

```java
import org.testng.annotations.AfterClass;
import org.testng.annotations.BeforeTest;
import org.testng.annotations.AfterTest;
import org.testng.annotations.BeforeSuite;
import org.testng.annotations.AfterSuite;

public class CalculateTest {
  @Test(dataProvider = "dp", expectedExceptions = ArithmeticException.class)
  public void f(int n, int s) {
      Calculate calc = new Calculate();
      int result;
      result = calc.add(n, s);
  }

  @DataProvider
  public Object[][] dp() {
    return new Object[][] {
      new Object[] { 4, 2147483647},
    };
  }
}
```

运行结果如图 5-42 所示，提示本应抛出异常而未抛出。

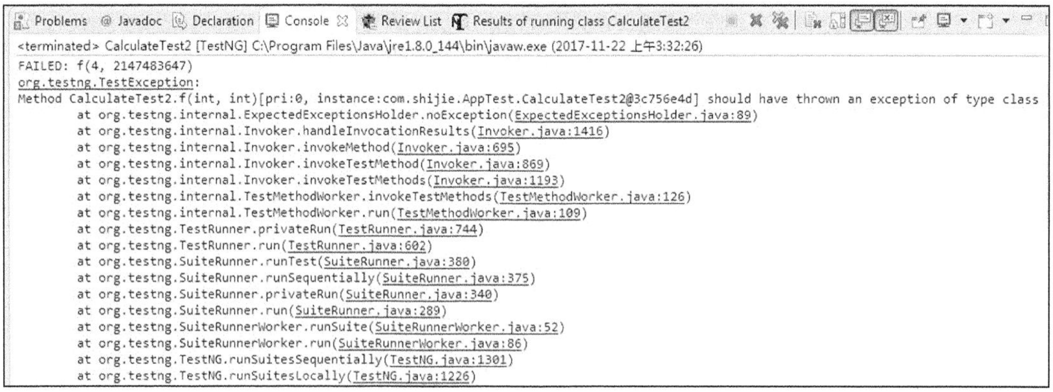

图 5-42　异常测试运行结果

5.3.10 测试报告

除了在 Eclipse 中展示外，TestNG 测试结果还自带了 HTML 版本的测试报告。执行完测试用例之后，会在项目的 test-output（默认目录）下生成测试报告，如图 5-43 所示。

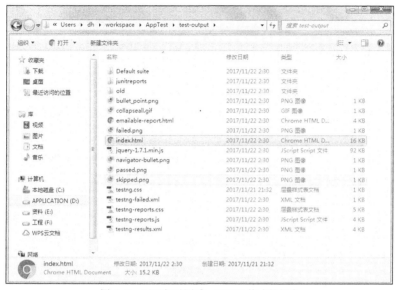

图 5-43　对于超时测试，HTML 格式的报告

打开 index.html 即可查看 TestNG 类型的测试报告，如图 5-44 所示。

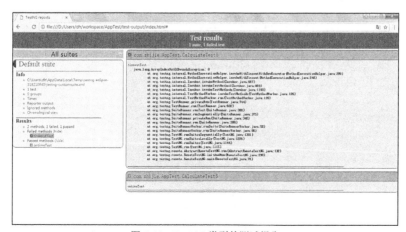

图 5-44　TestNG 类型的测试报告

5.3.11 断言

TestNG 在校验测试结果时使用 Assertion 类。Assertion 类的常用方法见表 5-7。

表 5-7 Assertion 类的常用方法

方法	描述
fail	直接失败测试用例,可以抛出异常
assertEquals	判断是否相等
assertFalse	判断是否为假
assertNotEquals	判断是否不相等
assertNotNull	判断是否不为空
assertNotSame	判断引用地址是否不相等
assertNull	判断是否为空
assertSame	判断引用地址是否相等
assertTrue	判断是否为真

修改测试类,加入 Assertion 判断 add 方法的计算结果是否与 expect 值一致。若一致,则测试用例运行成功;若不一致,则表明测试用例运行失败。具体代码如代码清单 5-16 所示。

代码清单 5-16   断言测试类

```java
package com.shijie.AppTest;

import org.testng.Assert;
import org.testng.annotations.Test;
import org.testng.annotations.BeforeMethod;
import org.testng.annotations.AfterMethod;
import org.testng.annotations.DataProvider;
import org.testng.annotations.BeforeClass;
import org.testng.annotations.AfterClass;
import org.testng.annotations.BeforeTest;
import org.testng.annotations.AfterTest;
import org.testng.annotations.BeforeSuite;
import org.testng.annotations.AfterSuite;


public class CalculateTest {
```

```
@Test(dataProvider = "dp")
public void f(int n, int s, int expected) {
    Calculate calc = new Calculate();
    int result;
    result = calc.add(n, s);
    Assert.assertEquals(result, expected);
}

@DataProvider
public Object[][] dp() {
  return new Object[][] {
    new Object[] { 1, 2, 3},
    new Object[] { 2, 999, 1000},
  };
}
```

在代码清单 5-16 中，数据传递者传参的时候加入了期望值（expected），用于在 Assertion 中进行比较。1 与 2 相加的期望值为 3，实际结果为 3。因为期望值和实际结果相等，所以测试用例运行成功。2 与 999 相加的期望值为 1000，实际结果为 1001。因为期望值和实际结果不相等，所以测试用例运行失败。运行结果如图 5-45 所示。

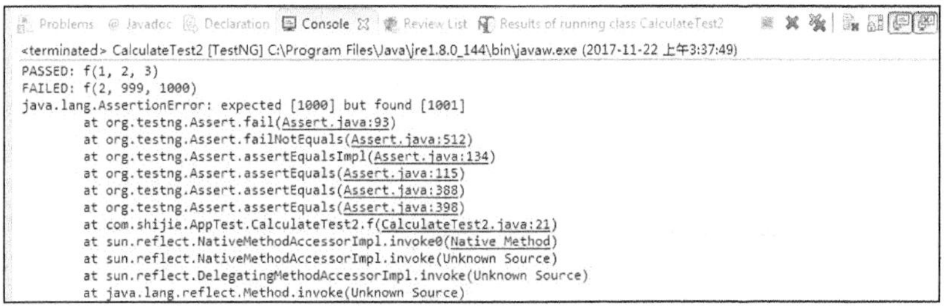

图 5-45 断言测试运行结果

5.3.12 通过 Maven 执行 TestNG 测试用例

以 TestNG 的测试方式在 IDE 中运行很方便。然而，这种方式不能解决依赖项和版本冲突的问题。Maven 解决了各种依赖问题，但是无法解决测试用例的忽略测试、依赖测试等运行方式设置方面的问题。只有将 Maven 和 TestNG 进行深度的捆绑，才能轻松解决各种问题。

如何通过 Maven 运行 TestNG 测试用例呢？Maven 本身并不是测试框架，在构建的时候需要借助插件来执行 TestNG 用例。该插件名为 maven-surefire-plugin。默认情况下，这个插件会自动扫描源码路径，即 src/test/java 下所有符合命名模式的测试用例。测试用例命名规则如下。

- 以 Test 开头的测试用例。
- 以 Test 结尾的测试用例。
- 以 TestCase 结尾的测试用例。

利用该插件，用户可以执行跳过测试、忽略测试等 TestNG 特性的测试。也可以通过 <suiteXmlFiles>标签指定 TestNG 配置 XML 文件（如 testng.xml，文件名任意取，在这里指定即可）。在 5.3 节中，所有 TestNG 测试用例的运行配置文件都为 testng.xml，所以代码清单 5-17 中设置 TestNG 的运行配置文件为 testng.xml。

代码清单 5-17　　　　　　　　　通过 Maven 配置 TestNG

```xml
<piugins>
 <plugin>
        <groupId>org.apache.maven.plugins</groupId>
        <artifactId>maven-surefire-plugin</artifactId>
        <version>2.4</version>
        <configuration>
            <!-- 解决用 Maven 执行 test 时日志乱码的问题-->
            <argLine>-Dfile.encoding=UTF-8</argLine>
            <!-- 解决 Maven 内存溢出的问题 -->
            <argLine>-Xms1024m -Xmx1024m -XX:PermSize=128m -XX:MaxPermSize= 128m</argLine>
            <forkMode>never</forkMode>
            <suiteXmlFiles>
            <suiteXmlFile>testng.xml</suiteXmlFile>
            </suiteXmlFiles>
            <!--定义 Maven 运行测试生成的报表路径 -->
            <reportsDirectory>./result/test-report</reportsDirectory>
        </configuration>
```

```
</plugin>
…….
</plugins>
```

其中,几个选项的作用如下。

- plugin:用来配置插件。
- configuration:用来扩展配置项。
- argLine:用来配置 JVM 参数和语言格式。
- forkMode:指明是要为每个测试创建一个进程,还是所有测试在同一个进程中完成。
- suiteXmlFiles:指定 TestNG 的运行配置文件。

TestNG 的 forkMode 属性见表 5-8。

表 5-8 TestNG 的 forkMode 属性

forkMode 属性的值	描述
once	在一个进程中完成所有测试。once 为默认设置
always	在一个进程中并行运行的脚本,surefire 版本要在 2.6 以上才提供这个功能
pretest	为每一个测试创建一个新进程,为每个测试创建新的 JVM 是彻底进行单独测试的方式,但也是最慢的
never	不创建新进程

TestNG 默认执行 test 包下的*Test.java 和 Test*.java 以及*TestCase.java。在 Pom.xml 中添加 TestNG 依赖就能执行 TestNG 测试。

5.4　Log4j 日志

自动化测试中,另一个必要功能便是执行结果的收集汇报模块。长时间运行测试用例后,能自动生成一份详细的测试报告以方便测试人员查看问题并解决问题。这是自动化测试生命流程的最后一个环节,若缺少日志功能,前面的工作将会事倍功半。

Log4j 是一个开源项目。在自动化框架中通过使用 Log4j,可以控制日志信息输送的目的地是控制台、文件和 GUI 组件等。它的配置文件用来设置日志级别,文件格式可以为键值对或者 XML 格式。

5.4.1　Log4j 安装

Log4j 的下载地址为清华大学开源镜像网，如图 5-46 所示。

图 5-46　Log4j 安装地址

选择其中一个镜像，下载最新版本 1.2.17，如图 5-47 所示。

图 5-47　选择 Log4j 版本

解压下载的 ZIP 包。将解压后的 log4j-1.2.17.jar 添加到 Eclipse 的 Build Path 中，如图 5-48 所示。

图 5-48　Log4j 的 Jar 包所在位置

右击 AppTest 项目，从上下文菜单中选择 Build Path→Configure Build Path。在弹出的 Properties for AppTest 对话框中，选择 Libraries 选项卡，单击 Add External JARs 按钮，选择 log4j-1.2.17.jar 所在的位置，回到 Properties for AppTest 对话框，单击 OK 按钮，Log4j 的配置完成，如图 5-49 所示。

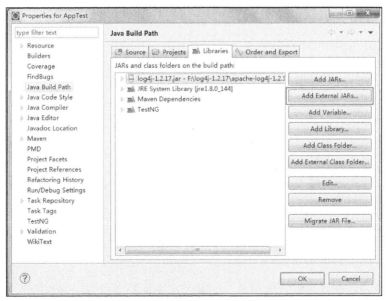

图 5-49　完成 Log4j 的配置

配置成功后，AppTest 项目的目录下会出现 Referenced Libraries 目录，如图 5-50 所示。

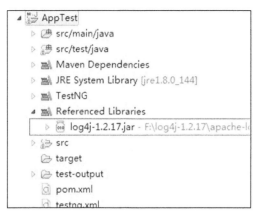

图 5-50　Log4j 成功配置到测试项目中

5.4.2　Log4j 配置文件

首先，在 AppTest 项目的根目录下创建 log4j.properties。

在 AppTest 项目的目录中右击，选择 New→File，弹出 New 对话框，输入文件名 log4j.properties，如图 5-51 所示。

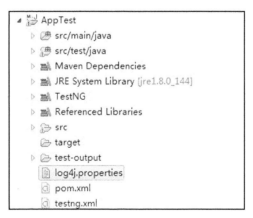

图 5-51　创建 Log4j 配置文件

Log4j 可以通过配置文件灵活地进行配置，而不需要修改应用的代码。另外，配置文件只能采用 XML、json、yaml 以及 properties 格式。本书采用 properties 格式。

接下来，配置 log4j.properties 文件。

下面是一个 log4j.properties 文件的完整配置，在这个文件（代码清单 5-18）中进行了详细的注释。

代码清单 5-18　　　　　　　　　　Log4j 配置文件

```
log4j.rootLogger=info, toConsole, toFile
#此句将等级为 info 的日志信息输出到 console 和 file 的两个目的地 toConsole 和 toFile
#名字可以随意命名
log4j.appender.file.encoding=UTF-8
#它可以使用任何字符编码。默认情况下是特定于平台的编码方案
log4j.appender.toConsole=org.apache.log4j.ConsoleAppender
#定义 toConsole 的输出类型。ConsoleAppender 代表控制台
log4j.appender.toConsole.Target = System.out
log4j.appender.toConsole.layout=org.apache.log4j.PatternLayout
#定义 toConsle 输出端的 layout 类型，PatternLayout 用于灵活地指定布局模块
log4j.appender.toConsole.layout.ConversionPattern=[%d{yyyy-MM-dd HH:mm:ss}] [%p] %m%n
#若使用 Pattern 布局，就要指定输出信息的具体格式
log4j.appender.toFile=org.apache.log4j.DailyRollingFileAppender
#定义 toFile 的输出类型。DailyRollingFileAppender 每天产生一个日志文件
log4j.appender.toFile.file=result/log/testlog.log
#定义 toFile 的输出端的文件名
log4j.appender.toFile.append=false
log4j.appender.toFile.Threshold =info
#Appender 可以有与之独立的记录器级别相关联的级别阈值水平。Appender 忽略具有级别低于阈值级别的任何日志消息
log4j.appender.toFile.layout=org.apache.log4j.PatternLayout
log4j.appender.toFile.layout.ConversionPattern=[%d{yyyy-MM-dd HH:mm:ss}] [%p] %m%n
```

针对代码清单 5-18，下面详细讲述每个字段的含义。

（1）Log4j 的配置文件需要类型不同，需要命名为 log4j-test 或者 log4j。其优先级如下。

- log4j-test.properties
- log4j-test.yaml 或者 log4j-test.yml
- log4j-test.json 或者 log4j-test.jsn
- log4j-test.xml

- log4j.properties
- log4j.yaml 或者 log4j-.yml
- log4j.json 或者 log4j.jsn
- log4j.xml

(2) Appenders 的配置见表 5-9。

表 5-9　Appenders 的配置

Appenders 的配置项	描　　述
ConsoleAppender	控制台，输出结果到 system.out 或者 system.err
FileAppender	输出结果到指定文件，同时可以指定输出数据的格式，append=true 指定追加到文件末尾
DailyRollingFileAppender	每天产生一个日志文件
RollingFileAppender	文件大小到达指定数值后生成新文件
WriterAppender	将日志信息以流格式发送到任意指定的地方

 这些节点的名称不能随便修改。

(3) Level 的设置见表 5-10。

表 5-10　Level 的设置

Level 的值	描　　述
OFF	关闭所有日志
FATAL	致命错误
ERROR	严重错误
WARN	一般警告
INFO	一般要显示的信息
DEBUG	调试信息
TRACE	追踪信息
ALL	输出所有日志

优先级从高到低为 FATAL>ERROR>WARN>INFO>DEBUG>TRACE。通过定义级别，可以控制程序中相应级别的日志信息输出等级。只有大于或者等于这个级别的日志才会输出。如 Log4j 默认的优先级为 ERROR，只有大于或者等于这个级别的日志（FATAL、ERROR）能正常输出，WARN、INFO、DEBUG 和 TRACE 日志将会被忽略。OFF 表示关闭所有的日

志信息，ALL 表示输出所有的日志信息。

（4）Layout 的配置见表 5-11。

表 5-11　Layout 的配置

Layout 的配置项	描述
HTMLLayout	以 HTML 表格形式布局
PatternLayout	可灵活地指定布局模式
SimpleLayout	包含日志信息的级别和信息字符串
TTCCLayout	包含日志产生的时间、线程、类别等信息

（5）ConversionPattern 的配置见表 5-12。

表 5-12　ConversionPattern 的配置

ConversionPattern 的配置项	描述
%m	输出代码中指定的消息
%p	输出优先级，即 DEBUG、INFO、WARN、ERROR 和 FATAL
%r	输出自应用启动到输出该 log 信息耗费的毫秒数
%c	输出所属的类目，通常就是所在类的全名
%t	输出产生该日志事件的线程名
%n	输出一个回车换行符，Windows 平台中为"rn"，UNIX 平台中为"n"
%d	输出日志时间点的日期或时间，默认格式为 ISO8601，也可以在其后指定格式，比如%d{yyyy MMM dd HH:mm:ss,SSS}，输出类似 2002 年 10 月 18 日 22：10：28，921
%l	输出日志事件的发生位置，包括类目名、发生的线程，以及在代码中的行数

5.4.3　Log4j 引用

要在某个类中使用 Log4j 记录日志，首先要声明一个成员变量，然后调用其方法即可。新建一个类（TestLog4j），如代码清单 5-19 所示，测试 Log4j 是否配置成功。

代码清单 5-19　　　　　　　　　Log4j 示例

```
import org.apache.log4j.Logger;
import org.apache.log4j.PropertyConfigurator;
public class TestLog4j {

static Logger log = Logger.getLogger(TestLog4j.class.getName());
```

```
public static void main(String[] args){
        PropertyConfigurator.configure("log4j.properties");
        // 如何处理log4j.xml文件
        DOMConfigurator.configure("F://book//AppTest//config//log4j.xml");
        log.debug("Hello this is an debug message");
        log.info("Hello this is an info message");
        log.error("Hello this is an error message");


}
}
```

运行脚本，若看到控制台以及 result/log 目录下有日志输出，则说明配置成功。在配置文件代码清单 5-19 中设置输出 info 级别以上的日志，因为代码中 "Hello this is an debug message" 这句日志为 debug 日志，所以它并未输出。控制台日志如图 5-52 所示。

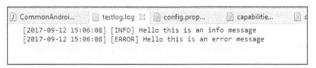

图 5-52　控制台日志

日志文件如图 5-53 所示。

图 5-53　日志文件

> 若将 Log4j 配置文件放置在 src 目录下，在调用的时候，不用导入文件路径，在程序执行的时候，会自动搜索路径。如果切换文件地址，需要手动配置文件加载的方法。例如，上例中使用的是相对路径：PropertyConfigurator.configure("log4j.properties")。也可以使用绝对路径：PropertyConfigurator.configure ("F://book//AppTest//log4j.properties")。否则会提示 log4j:WARN No appenders could be found for logger 的错误。配置 config 文件夹，在项目的目录下右击，从上下文菜单中选择 Build Path，在里面进行配置即可。

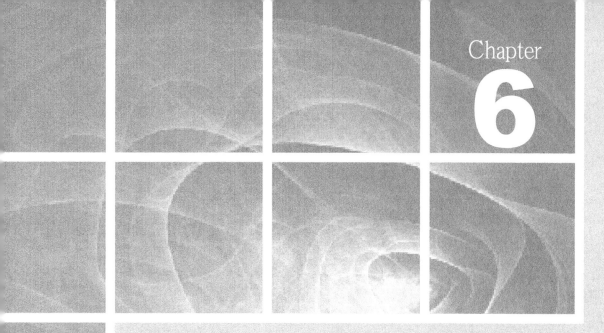

第 6 章

Android 自动化测试基础精讲

第 6 章 Android 自动化测试基础精讲

在学习了自动化测试项目管理工具（Maven）和 Java 单元测试框架（TestNG）后，本章将正式进入 Android 自动化测试的讲解。

adb 是 PC 端控制移动端的桥梁，也是移动端自动化测试（包括 Appium 自动化测试）的基础，本章会重点介绍它。

自动化功能测试是以程序测试程序，以代码代替思维，以脚本的运行代替手工操作。本章中，会结合手工测试介绍 App 测试的常见手工操作、按键以及坐标点获取，为后面书写自动化测试用例代码打下基础。

除此之外，本章的侧重点为 Appium 自动化测试工具，以及完成 Appium 配置所需要的前期准备工作，如获取 Appium 配置项（如获取 Main Activity 值）等。

6.1 adb 命令

adb 的全称为 Android Debug Bridge，它是一种命令行工具，在 Android 设备与 PC 之间起到调试桥梁的作用，方便用户通过 PC 直接操作 Android 设备。在我们的平时工作生活中常见的一些通过计算机操作手机的软件（如 360 手机助手、豌豆荚等）都通过 adb 实现对手机的操控。本节介绍一些常用的 adb 命令和测试中要使用到的 adb 命令。

6.1.1 在手机上启动 USB 调试

使用 adb 命令控制 Android 设备之前，需要先启动 Android 设备上的 USB 调试。以 Oppo 手机为例，执行以下步骤来完成 USB 调试的启动。

在手机主界面中，选择"设置"→"常规"→"关于手机"→"版本号"，连续单击"版本号"7 次，会提示进入"开发者选项"界面，如图 6-1 所示。

在"开发者选项"界面中，单击右上角的滑动按钮，开启 USB 调试，如图 6-1 所示。在弹出的对话框上，单击"开启"按钮，如图 6-2 所示。

用数据线连接计算机和手机，会提示"允许 USB 调试吗"，如图 6-3 所示。

图 6-1 进入"开发者选项"界面

图 6-2　开启 USB 调试

图 6-3　允许 USB 调试

勾选"一律允许使用这台计算机进行调试"复选框，单击"确定"按钮，启动手机调试。

6.1.2　adb 命令环境搭建

通常，对于已经安装 Android SDK 的设备，adb 工具位于 Android SDK 安装目录的 platform-tools 文件夹下。若安装 SDK 的时候将该目录配置到了环境变量里，则可以直接在 Windows 命令行工具中使用；若没有加入系统环境变量中，则建议加入，以方便使用。platform-tools 目录的结构如图 6-4 所示。

图 6-4　adb 在 platform-tools 目录中的位置

如果安装了 SDK 并配置了相关环境变量，使用命令行窗口来调用 adb 命令。首先打开"开始"菜单，单击"运行"，然后在文本框中输入"cmd"，按<Enter>键，就直接进入 Windows 的命令行界面，如图 6-5 所示。

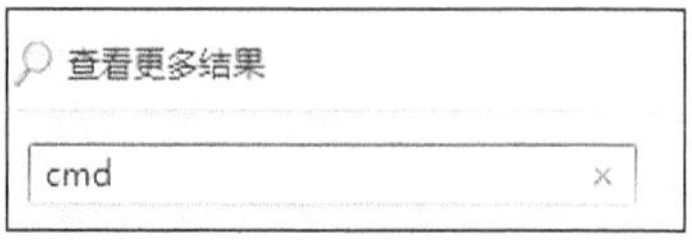

在命令行界面中，输入"adb"，如果安装 adb 成功，则会回传 adb 的调试信息。如图 6-6 所示，此时表示 adb 已经可以正常使用了，用户可以在命令行工具中直接输入 adb 命令来完成各种操作。

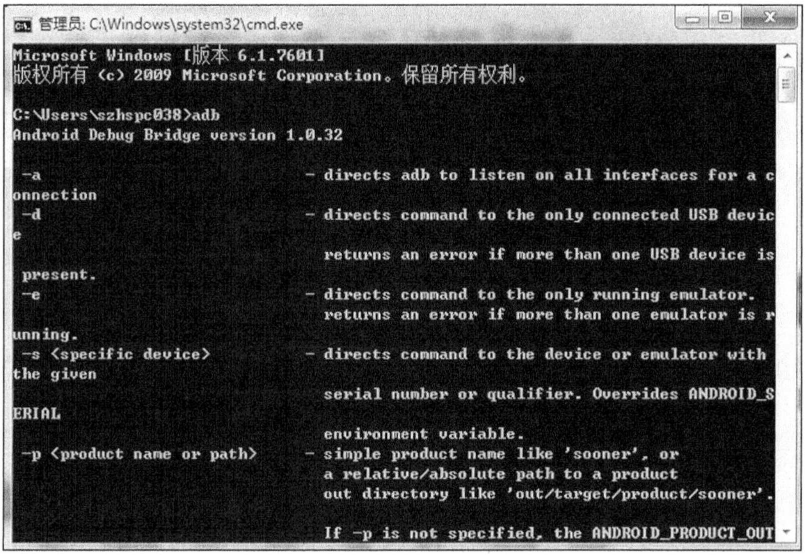

图 6-5　进入 Windows 命令行命界

图 6-6　检查 adb 是否安装成功

6.1.3　adb 组织结构简介

adb 的组织结构如图 6-7 所示。

adb 主要由以下 3 部分组成。

- 运行在 PC 端的客户端，可以通过 cmd 命令行使用 adb 命令启动客户端（借助其他一些工具（如手机助手、豌豆荚等）也可以直接启动 adb 客户端）。当启动一个 adb 客户端的时候，客户端首先确认是否已经存在正在运行的 adb 进程，如果没有则启动进程。
- 作为后台进程运行的服务器端，adb 服务器运行后会自动绑定本地的 TCP 端口 5037，监听客户端发来的命令。所有的 adb 客户端都通过端口 5037 与 adb 服务器进行对话。

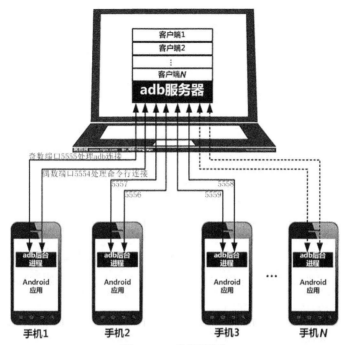

图 6-7 adb 的组织结构

- 在 Android 设备上，以后台进程的形式运行的 adb 后台守护进程[①]。

adb 服务器运行后，自动绑定本地的 TCP 端口 5037，监听所有通过 5037 端口发来的客户端请求。接着 adb 服务器会扫描 5555～5585 范围内所有的奇数端口来定位设备实例。一旦找到运行在 Android 设备上的 adb 守护进程，就建立一个基于该端口的连接。

 任何 Android 设备或者模拟器设备都会取得两个连续的端口，即一个奇数端口和一个偶数端口。奇数端口用来处理 adb 连接，偶数端口用来处理控制台的连接。

6.1.4 adb 常用命令

讲述了 adb 技术的组织结构之后，下面就介绍常用的 adb 命令。首先，如前所述，将手

---

① 守护进程（Daemon），通常在系统后台运行，没有控制终端且不与前台交互。一般作为系统服务使用。

机连接到 PC，通过命令行启动 adb 进程。

1. adb 调试命令

1）adb devices

因为 adb 可以同时管理多个 Android 设备，所以 adb 命令的一般形式如下。

```
adb [-e|-d|-s<设备序列号>]子命令
```

如果设备只有一个模拟器实例或者一台 Android 设备，可以省略-e、-d、-s 等选项。

如果设备有多台模拟器实例和多台 Android 设备，使用-s 来区分要执行 adb 子命令的设备或者模拟器。

如果设备只有一个模拟器实例和一台 Android 设备，可以通过-e（emulator）或者-d（device）来选择是在模拟器还是 Android 设备上执行 adb 子命令。

图 6-8　adb devices 命令的执行结果

adb devices 命令的执行结果如图 6-8 所示。

执行结果中部分选项的含义如下。

- d300be9d：序列号，adb 创建的字符串，通过它唯一地识别一个模拟器/设备实例。
- device：状态。offline 表示没有与 adb 相连接或者无法响应；no device 表示没有模拟器/设备运行；device 表示有模拟器/设备运行。

> -s 后需要跟设备序列号，可以通过 "adb devices" 获得。

2）adb forward

要发布端口，命令如下。

```
adb forward tcp:####tcp:####
```

用户可以设置任意的端口号，作为主机向模拟器或 Android 设备的请求端口。一个模拟器/设备实例的某一特定主机端口向另一个端口转发请求。下面演示了如何实现从主机端口 5555 到模拟器/设备端口 8000 的请求转发。

```
adb forward tcp:5555 tcp:8000
```

这样所有发往 5555 端口的数据都会被转发到 8000 端口上。

3）adb kill-server

要停止 adb 服务器，命令为 kill-server，如图 6-9 所示。

在某些情况下，用户可能需要终止 Android 调试系统的运行，然后再重新启动它。例如，如果 Android 调试系统不响应命令，则可以先终止服务器，然后再重启，这样就可能解决这个问题。

图 6-9　通过 adb 关闭进程

4）adb start-server

要启动 adb 服务器，命令为 start-server，如图 6-10 所示。

图 6-10　通过 adb 启动服务器

2. adb 连接命令

1）adb connect

要连接某个设备，常用命令如下。

```
adb connect <host>[:<port>]
```

如果在测试过程中 PC 的 USB 端口有限，不可能一直使用 USB 模式测试 Android 设备，则可以通过 Wi-Fi 的形式完成 PC 和 Android 设备的连接。

前提条件是 PC 和手机都已连接同一个 Wi-Fi，并且手机通过 USB 和 PC 相连。

接下来，执行如下步骤。

（1）输入"adb devices"命令验证 PC 通过 USB 模式成功连接手机设备，如图 6-11 所示。

（2）在 tcpip 模式下重启 adb 端口 5555。

```
adb tcpip 5555
```

执行结果如图 6-12 所示。

```
C:\Users\dh>adb devices
List of devices attached
* daemon not running. starting it now at tcp:5037 *
* daemon started successfully *
c1aeae297d72    device

C:\Users\dh>
```

图 6-11　验证 PC 通过 USB 成功连接手机设备

```
C:\Users\dh>adb tcpip 5555
restarting in TCP mode port: 5555
```

图 6-12　重启端口

（3）查找手机的 IP 地址。以 Oppo 手机为例，选择"设置"→"关于手机"→"状态消息"→"IP 地址"。这里 IP 地址设置为 192.168.1.123，如图 6-13 所示。

图 6-13　在手机上查看 IP 地址

（4）连接到手机 IP 地址。

```
adb connect 192.168.1.123
```

连接结果如图 6-14 所示。

```
C:\Users\dh>adb connect 192.168.1.123
connected to 192.168.1.123:5555
```

图 6-14　连接指定 IP 地址

（5）输入命令"adb devices"验证是否成功连接手机和 PC，如图 6-15 所示。

```
C:\Users\dh>adb devices
List of devices attached
192.168.1.123:5555      device
```

图 6-15　验证通过 Wi-Fi 是否成功连接手机和 PC

（6）通过以下命令断开连接。

```
adb disconnect 192.168.1.123
```

执行结果如图 6-16 所示。

```
C:\Users\dh>adb disconnect 192.168.1.123
disconnected 192.168.1.123
```

图 6-16　断开连接

2）adb USB

在 USB 连接模式下重启 adb 服务器。

3．adb 包管理命令

1）adb install

要安装软件，命令的一般形式如下。

```
adb install <apk 文件路径>
```

install 命令要求指定所要安装的 .apk 文件的路径。如果要将 com.baidu.searchbox_38032640.apk 安装到移动设备上，则需要指定 apk 文件所在位置的全路径。提示信息 Success 表示安装成功，如图 6-17 所示。

```
C:\Users\dh>adb install F:\com.baidu.searchbox_38032640.apk
F:\com.baidu.searchbox_38032640.apk: 1.... 3.4 MB/s (51039331 bytes in 14.467s)
        pkg: /data/local/tmp/com.baidu.searchbox_38032640.apk
Success
```

图 6-17　adb install 命令

2）adb uninstall

要卸载软件，命令的一般形式如下。

```
adb uninstall <软件包名>
adb uninstall -k <软件包名>
```

这里，-k 参数表示卸载软件但是保留配置和缓存文件。要卸载 com.baidu.input_miv6 包，命令如图 6-18 所示。

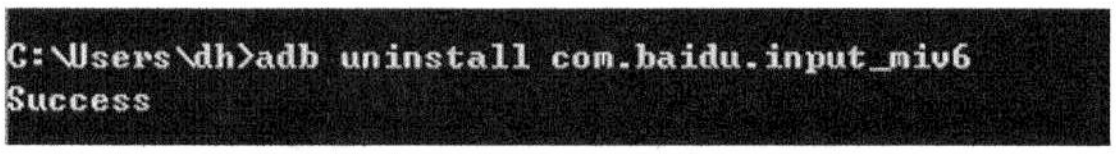

```
C:\Users\dh>adb uninstall com.baidu.input_miv6
Success
```

图 6-18 通过 adb 卸载软件

卸载软件需要指明软件的包名。软件包名可以使用 adb logcat 命令获取，后面章节会详细描述相关内容。

3）adb shell pm

命令 pm 的全称是 package manager，可以使用 pm 命令执行应用的安装并查询一些安装包的常用信息。以下命令用于列出所有安装的包信息。

```
adb shell pm list packages
```

执行结果如图 6-19 所示，读者显示的信息可能和本书有差异，因为不同的手机或模拟器环境下的结果可能不同。

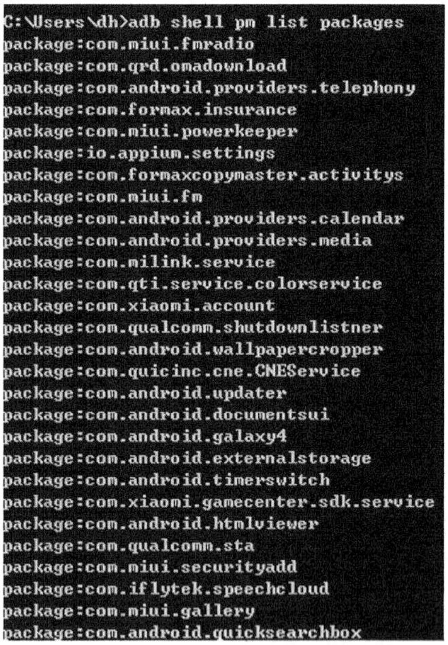

```
C:\Users\dh>adb shell pm list packages
package:com.miui.fmradio
package:com.qrd.omadownload
package:com.android.providers.telephony
package:com.formax.insurance
package:com.miui.powerkeeper
package:io.appium.settings
package:com.formaxcopymaster.activitys
package:com.miui.fm
package:com.android.providers.calendar
package:com.android.providers.media
package:com.milink.service
package:com.qti.service.colorservice
package:com.xiaomi.account
package:com.qualcomm.shutdownlistner
package:com.android.wallpapercropper
package:com.quicinc.cne.CNEService
package:com.android.updater
package:com.android.documentsui
package:com.android.galaxy4
package:com.android.externalstorage
package:com.android.timerswitch
package:com.xiaomi.gamecenter.sdk.service
package:com.android.htmlviewer
package:com.qualcomm.sta
package:com.miui.securityadd
package:com.iflytek.speechcloud
package:com.miui.gallery
package:com.android.quicksearchbox
```

图 6-19 通过 pm 获取已安装包名

其中参数描述如下所示。

-f：显示每个包的文件位置。

-d：使用过滤器，只显示禁用应用的包名。

-e：使用过滤器，只显示可用应用的包名。

-s：使用过滤器，只显示系统应用的包名。

-3：使用过滤器，只显示第三方应用的包名。

-i：查看应用的安装者。

例如，要查看第三方应用的包名，命令如下。

```
adb shell pm list packages -3
```

执行结果如图 6-20 所示。

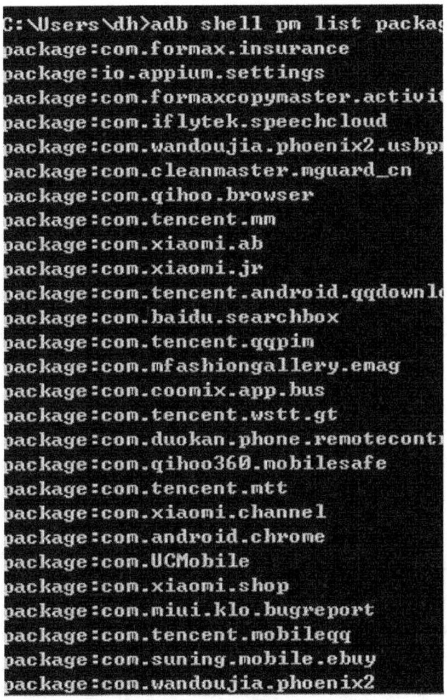

图 6-20　通过 pm 获取第三方应用的包名

4．adb 文件管理命令

1）adb push

要从计算机上发送文件到设备上，命令如下。

```
adb push <本地路径><远程路径>
```

用 push 命令可以把计算机上的文件或者文件夹复制到设备（手机）中。以下命令将本机文件"F:\blacklist.txt"传输到设备上的"/data/local/tmp"目录下。

```
adb push F:\blacklist.txt /data/local/tmp
```

例如在日常生活中，为了将计算机上的照片传输到手机中，可执行 adb push 命令，运行结果如图 6-21 所示。

```
C:\Users\szhspc038>adb push E:\car.jpg /storage/emulated/legacy/DCIM/Camera/
5007 KB/s (297424 bytes in 0.058s)
```

图 6-21 通过 adb push 传输文件

 传输文件之前，需要查看待传入文件夹是否有可写权限。

2）adb pull

要从设备上下载文件到计算机，命令如下。

```
adb pull <远程路径> <本地路径>
```

用 pull 命令可以把设备(手机)上的文件复制到计算机中。例如，以下命令将 window_dump.xml 复制到计算机的 F 盘下。

```
adb pull /storage/emulated/legacy/window_dump.xml F:\
```

在日常生活中，当将照片从手机中导入计算机时，首先要找到照片所存放的路径 /storage/emulated/legacy/DCIM/Camera/，然后执行 adb pull 命令，选择一张照片，传输到 E 盘中，如图 6-22 所示。

```
C:\Users\szhspc038>adb pull /storage/emulated/legacy/DCIM/Camera/IMG_20160829_1
3401.jpg  E:\
5715 KB/s (99494 bytes in 0.017s)
```

图 6-22 通过 adb pull 下载文件

5．adb 日志命令

1）adb logcat

要查看日志，命令如下。

```
[adb] logcat [<option>] ... [<filter-spec>] ...
```

一般来说，无线通信的日志非常多，在运行时没必要去全部记录，但还是可以通过以下命令获取想要的日志。

```
adb logcat
```

执行结果如图 6-23 所示。

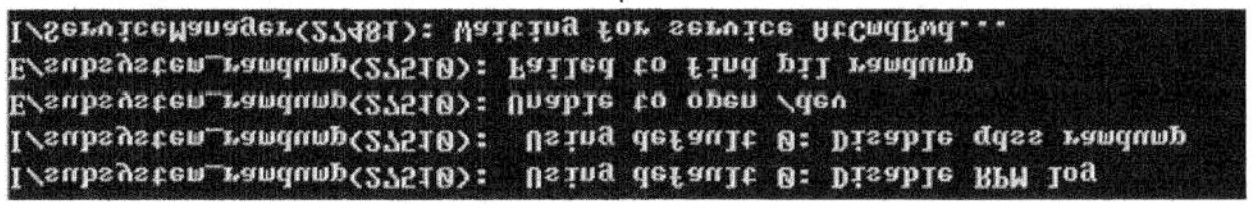

图 6-23 通过 adb logcat 获取日志信息

每一行的首字母 I、E 等表示日志的级别,"/"与":"之间的内容表示消息的来源。常见的日志级别包括 V、D、I、W、E,见表 6-1。

表 6-1 adb logcat 日志的级别

日 志 级 别	描　述
V	表示冗余级别的日志信息
D	表示调试级别的日志信息
I	表示信息级别的日志信息
W	表示警告级别的日志信息
E	表示错误级别的日志信息

其中,由上到下级别越来越高。可以加上过滤器对于输出日志进行过滤,高于过滤器设置的级别都会显示出来。

设置过滤器的命令如下。

```
adb logcat*:I
```

这样 I、W、E 级别的日志都会显示出来。

有时候,如果日志太多,就需要使用以下命令,清除所有以前的日志。

```
adb logcat -c
```

当做 Android 前端性能测试的时候,为了获取应用的启动时间。可以打开 App,并执行如下命令。

```
adb logcat -d -s ActivityManager|findstr "Displayed"
```

执行结果如图 6-24 所示。

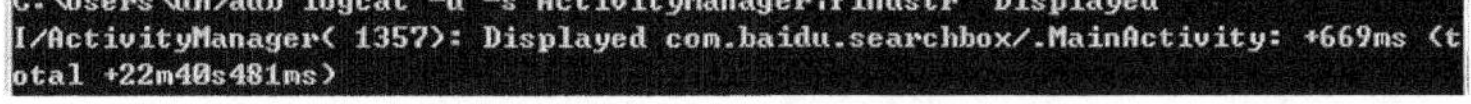

图 6-24 通过 adb logcat 获取日志信息

输入 adb logcat -d 命令,显示的日志就是刚刚操作那段时间内的日志,而且会自动退出

log 模式。

adb logcat -s XXX 命令是设置过滤用的。如果只想查看消息来源为 ActivityManager 的日志，就直接用 ActivityManager 替换最后的 XXX。

在卸载 App 或者针对具体的 App 执行命令的时候，需要获取 App 的包名。首先使用如下命令清空 logcat 日志信息。

```
adb logcat -c
```

-c 表示清空日志记录。然后，使用如下命令。

```
adb logcat ActivityManager:I *:s
```

接下来，在手机端启用待测 App，表示获取 Tag 为 ActivityManager 且输出级别大于 I 的日志。获取的日志，以及获取的包名 com.greenpoint.android.mc10086.activity，如图 6-25 所示。

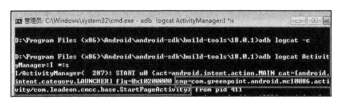

图 6-25　通过 adb logcat 获取包名和 Main Activity

2）adb shell dumpsys

Android 系统是基于 Linux 系统开发的，支持常见的 Linux 命令。可以通过"adb shell"指令进入设备或者模拟器 Shell 环境，也可以在"adb shell"后直接加上 Linux 指令及参数来完成指令的执行。命令的一般形式如下。

```
adb shell
adb shell [command]
```

在做 Android 前端性能测试的时候，为了检查是否有内存泄漏的情况，需要获取系统的内存值。

```
adb shell dumpsys meminfo <packagename>
```

packagename 为待测 APK 的包名，以百度 APK 为例。

```
adb shell dumpsys meminfo com.baidu.searchbox
```

指令会返回内存使用情况，由于显示结果过多，这里只列出部分信息，如图 6-26 所示。

```
** MEMINFO in pid 26125 [com.baidu.searchbox] **
                    Pss    Private   Private   Swapped    Heap     Heap      Heap
                   Total     Dirty    Clean      Dirty    Size    Alloc      Free

   Native Heap    14826    14724         0       956    20328    20031       296
   Dalvik Heap    26400    26176         0     15600    47863    40429      7434
  Dalvik Other     1517     1516         0         0
         Stack     1168     1168         0         0
         Ashmem     132      128         0         0
       Gfx dev     5774     5744         0         0
     Other dev       17        0        16         0
      .so mmap    27614     1976     23860      2616
     .apk mmap     1062        0       788         0
     .ttf mmap       66        0        32         0
     .dex mmap    13174        0      9928         0
     .oat mmap     5759        0      2960         0
     .art mmap     3051     1632       424       112
    Other mmap     1010        4       936         0
       Unknown    38503    38500         0        52
         TOTAL   140073    91568     38944     19336    68191    60460      7730
```

图 6-26　通过 adb shell 获取内存信息

在做 Android 前端性能测试的时候，为了检查 CPU 占用性率是否过高，需要获取 CPU 信息。

```
adb shell dumpsys cpuinfo < packagename >
```

Packagename 为待测 APK 的包名，以百度 APK 为例。

```
adb shell dumpsys cpuinfo com.baidu.searchbox
```

由于显示结果过多，这里只列出部分信息，如图 6-27 所示。

```
 0.2% 26306/com.tencent.mobileqq: 0.2% user + 0% kernel / faults: 16 minor 17 m
ajor
 0.2% 29558/com.qihoo360.mobilesafe:GuardService: 0.2% user + 0% kernel / fault
s: 14 minor
 0.1% 7/rcu_preempt: 0% user + 0.1% kernel
 0.1% 13707/adbd: 0% user + 0.1% kernel / faults: 46 minor
 0.1% 29045/kworker/u12:0: 0% user + 0.1% kernel
 0.1% 29811/com.tencent.mobileqq:MSF: 0.1% user + 0% kernel / faults: 36 minor
 2 major
 0.1% 30229/com.tencent.mm:push: 0% user + 0% kernel / faults: 105 minor
 0% 3/ksoftirqd/0: 0% user + 0% kernel
 0% 45/kworker/u13:0: 0% user + 0% kernel
 0% 7369/com.lbe.security.miui: 0% user + 0% kernel / faults: 28 minor
 0% 26125/com.baidu.searchbox: 0% user + 0% kernel / faults: 271 minor 2 major
 0% 29579/kworker/0:0: 0% user + 0% kernel
 0% 29663/kworker/1:2: 0% user + 0% kernel
 0% 31906/com.tencent.mm:tools: 0% user + 0% kernel / faults: 104 minor 96 majo
0.4% TOTAL: 0.1% user + 0.2% kernel + 0% iowait
```

图 6-27　通过 adb shell 获取 CPU 信息

在做前端性能测试的时候，如果需要关注前台耗电量和后台耗电量，则可以使用如下命令。

```
adb shell dumpsys battery
```

执行结果如图 6-28 所示。

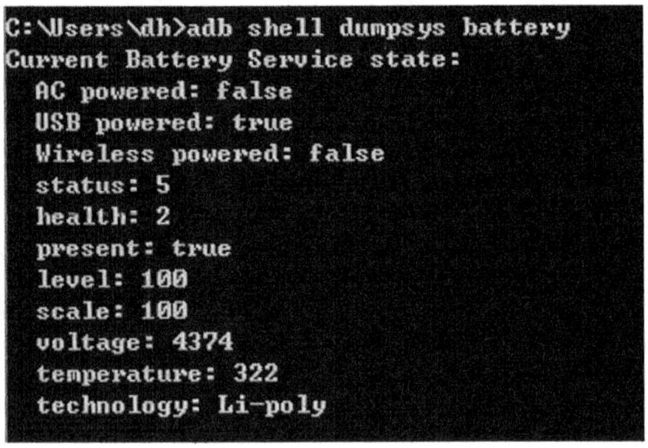

```
C:\Users\dh>adb shell dumpsys battery
Current Battery Service state:
  AC powered: false
  USB powered: true
  Wireless powered: false
  status: 5
  health: 2
  present: true
  level: 100
  scale: 100
  voltage: 4374
  temperature: 322
  technology: Li-poly
```

图 6-28 通过 adb shell 获取电量信息

dumpsys 是 Android 手机自带的调试工具，主要是用于转储当前 android 系统的一些信息。

6. adb 截图命令

截图命令如下。

```
adb shell screencap
```

例如，为了截取当前屏幕，保存到目录/sdcard/screen.png，并将截图上传到计算机，可以按以下步骤进行操作。

（1）输入以下命令进行截屏。

```
adb shell screencap /sdcard/screen.png
```

（2）执行如下命令，将截屏的图上传到 F 盘，目录为 F:\screenshot。

```
adb pull /sdcard/screen.png F:\screenshot
```

执行结果如图 6-29 所示。

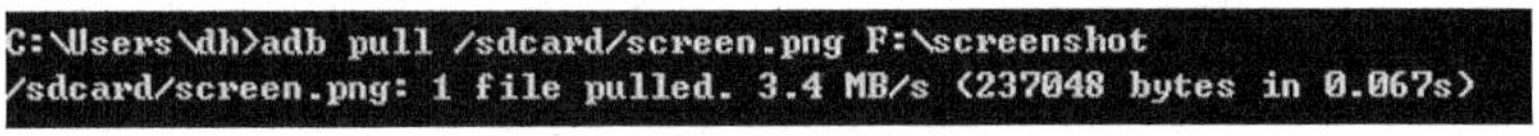

```
C:\Users\dh>adb pull /sdcard/screen.png F:\screenshot
/sdcard/screen.png: 1 file pulled. 3.4 MB/s (237048 bytes in 0.067s)
```

图 6-29 上传截图

7. adb 系统命令

```
adb get-product
```

要获取设备的 ID 和序列号，如图 6-30 所示，执行如下命令。

```
adb get-product
```

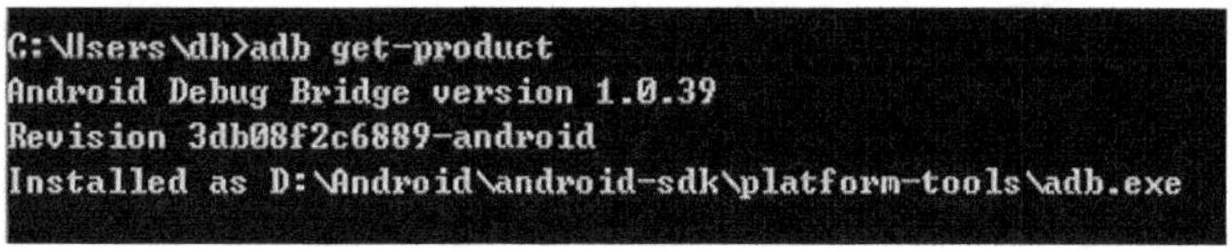

图 6-30　获取设备的 ID

要获取设备的序列号，如图 6-31 所示，执行如下命令。

```
adb get-serialno
```

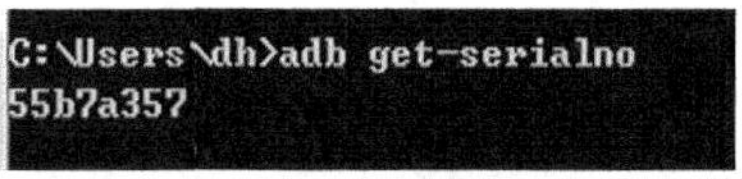

图 6-31　获取序列号

8．adb Activity 管理命令

```
adb shell am
```

am 命令的全称为 activity manager，用来执行不同的系统操作，可以通过此命令启动 Android 中的 Activity、Service（服务）和 Broadcast（广播进程）等组件。

如图 6-32 所示，通过 am 拨打电话，拨打的电话号码为 10086。

图 6-32　通过 am 拨打电话

其中，要注意以下两个选项。

- start：am 的子命令，因为是子命令，所以前面不用加短横线 "-"。它启动一个操作，后面为待启动的包名。
- -d：表示要操作的数据。

例如，使用 am 打开网页，如图 6-33 所示。

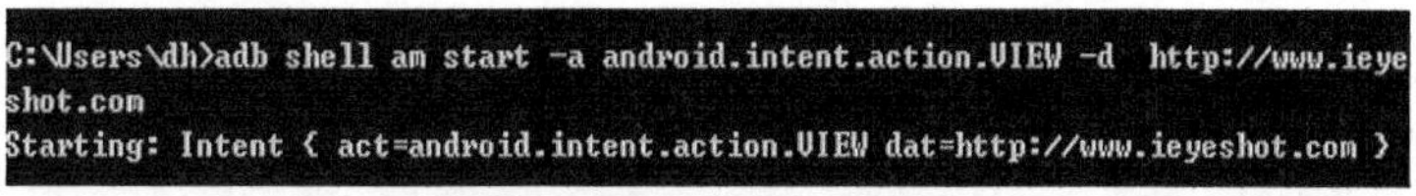

图 6-33　使用 am 打开网页

在测试的过程中，模拟电量低的情况，如图 6-34 所示。发出这个命令后，手机电量信息会迅速变为 3%，等一段时间后可以恢复正常电量值。

```
C:\Users\dh>adb shell am broadcast -a android.intent.action.BATTERY_CHANGED --ei "level" 3 --ei "scale" 100
Broadcasting: Intent { act=android.intent.action.BATTERY_CHANGED (has extras) }
Broadcast completed: result=0
```

图 6-34　通过 am 发送电量低的信息

6.1.5　adb 端口冲突问题解决

问题描述：端口冲突，报错如图 6-35 所示。

```
D:\09工具\adbgjb\adbgjb_xpgod\adb>adb devices
adb server is out of date.  killing...
ADB server didn't ACK
* failed to start daemon *
error: unknown host service
```

图 6-35　端口冲突报错

问题分析：adb 端口被其他应用程序进程占据，需要找到并关闭这个进程。

解决办法：使用命令 `netstat -ano | findstr "5037"`，获取端口占用信息，如图 6-36 所示。

```
C:\Users\dh>netstat -ano | findstr "5037"
  TCP    127.0.0.1:5037         0.0.0.0:0              LISTENING       4828
  TCP    127.0.0.1:5037         127.0.0.1:50868        TIME_WAIT       0
  TCP    127.0.0.1:5037         127.0.0.1:50869        TIME_WAIT       0
  TCP    127.0.0.1:5037         127.0.0.1:50870        TIME_WAIT       0
  TCP    127.0.0.1:5037         127.0.0.1:50871        TIME_WAIT       0
  TCP    127.0.0.1:5037         127.0.0.1:50873        TIME_WAIT       0
  TCP    127.0.0.1:5037         127.0.0.1:50874        TIME_WAIT       0
  TCP    127.0.0.1:5037         127.0.0.1:50893        TIME_WAIT       0
  TCP    127.0.0.1:5037         127.0.0.1:50894        TIME_WAIT       0
  TCP    127.0.0.1:5037         127.0.0.1:50951        TIME_WAIT
```

图 6-36　获取端口占用信息

其中，第 1 列表示连接协议；第 2 列表示本地 IP 和端口；第 3 列表示外部 IP 和端口；第 4 列表示连接状态；第 5 列表示进程的 PID。

关闭占据本地 PID 的进程，就是 PID 为 4828 的进程。

同时按下<Ctrl+Alt+Del>快捷键调出"Windows 任务管理器"窗口。在该窗口中，在菜单栏中单击"查看→选择列"。在弹出的"进程选择页列"对话框中，勾选"PID（进程标识符）"复选框，如图 6-37 所示。

图 6-37　设置在任务管理器中显示 PID 列

在"Windows 任务管理器"窗口中显示 PID 列，如图 6-38 所示。

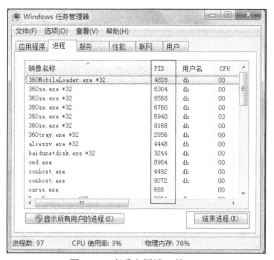

图 6-38　查看占用端口的 PID

单击"结束进程"按钮，再次使用 adb 命令基本可以解决问题。

6.2　Android 简介

在做移动端功能自动化测试时，用户需要执行代码去模拟手工操作，如单击图标、输入文本等。本节介绍了这些操作的含义。在写代码的时候，一些特殊场景（如测试用例需要减小手机音量等）下会使用到 Android 自带的物理返回键（Keycode），一些特殊场景（如被测对象无法通过对象的属性进行定位）下只能使用坐标点定位的方式来完成测试。本节会对这些物理返回键进行汇总，对 Android 坐标点的获取进行讲解，方便后面自动化测试的使用。

6.2.1　Android 常规动作

Android 系统中，与手势相关的常规动作如下。

1）单击

作为最常用手势，如图 6-39 所示，单击即轻触屏幕一下。它主要用于打开程序，以及功能表切换，如在"我的营业厅"中，通过单击"登录"按钮进入登录界面。

2）长按

长按，也叫点住、按住，指按住屏幕超过两秒。此动作通常用来调出菜单。如在主界面中，在空白处长按进入组件调整界面，对图标的位置进行编辑。当需要转发短信时，在短信对话界面长按短信内容，必然会弹出菜单，菜单中通常会有"转发"选项。在进行文字编辑时，长按文字部分，然后调出光标进行精准定位。当然，此动作也可用于多选、快捷视图。如在"图库（相册）"中，默认视图下，长按相册文件夹可对文件夹进行多选操作；打开相册文件夹后，长按照片同样可对照片进行多选操作，选择完毕后可进行批量操作（发送、删除）。效果如图 6-40 所示。

图 6-39　单击

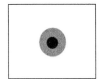
图 6-40　长按

3）拖动

准确来说，拖动应该叫作按住并拖动。拖动是主屏幕编辑时的常见动作，如对桌面小组件或者图标进行位置编辑等。另外，它也用于进度定位，如播放音乐或者视频时，常常需要拖动进度条。效果如图 6-41 所示。

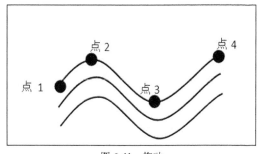

图 6-41　拖动

4）双击

双击就是短时间内连续双击屏幕两次，主要用于快速缩放。当浏览图片时，双击可快速放大，再次双击则复位；当浏览网页时，对文章正文部分双击可使文字自适应屏幕。当然，在某些视频播放器中双击可切换至全屏模式。效果如图 6-42 所示。

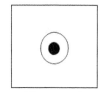

图 6-42　双击

5）双击拖动

放大或者缩小单击区域的内容，如单击松开后立刻再次按住屏幕上下拖动，向上拖动放大区域内容，向下拖动缩小区域内容。效果如图 6-43 所示。

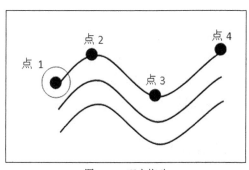

图 6-43　双击拖动

6）滑动

滑动，也是一个常见操作，主要用于查看屏幕无法完全显示的页面，功能类似于鼠标的滚轮。此操作主要用于查看图片、网页和纯文本。比如，在通讯录上滑动查找指定的人，在看书的时候可以通过从左到右或者从右到左实现页面的切换，在看照片的时候也可以通过滑动来切换照片。效果如图 6-44 所示。

7）放大

放大，是查看图片、网页时最常见的操作。照相时也可使用扩大手势来拉长焦距。如看短信的时候，可以通过放大来改变字体。效果如图 6-45 所示。

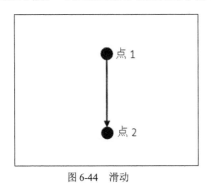
图 6-44　滑动

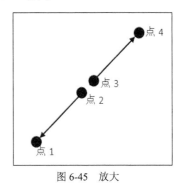
图 6-45　放大

8）缩小

缩小和放大相反，用两个手指按住，向相互接近的方向移动并抬起。查看照片时可用于缩小照片，照相时可用于拉短焦距。例如看短信的时候，可以通过缩小来缩小字体，如图 6-46 所示。

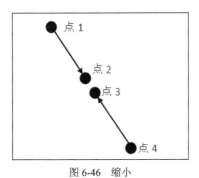
图 6-46　缩小

6.2.2　Android 的按键和 Keycode

每一种 Android 设备一般都会提供除字母和数字外的物理功能键，且根据不同的 Android 制造商而不同。如果这些功能键设计良好，对于正在开发的 App 应用，能给用户带来好的交互体验。在做功能自动化测试的过程中，也经常会需要调用这些 Keycode，手机中最常见的按键包括 HOME、MENU、BACK、VOLUME_UP、VOLUME_DOWN、RecentApps、POWER 和 Dpad 等。下面介绍一个标准的 Android 设备的功能按键。

1．电话键

电话键见表 6-2。

表 6-2 电话键

功 能 键	Keycode	描 述
KEYCODE_HOME	3	返回主菜单
KEYCODE_MENU	82	返回主菜单
KEYCODE_BACK	4	回到上一屏
KEYCODE_VOLUME_UP	24	增大音量
KEYCODE_VOLUME_DOWN	25	减小音量
KEYCODE_VOLUME_KEYCODE_MUTE	164	使扬声器静音
KEYCODE_POWER	26	打开设备或者唤醒设备
KEYCODE_SEARCH	84	打开一个搜索
KEYCODE_CAMARA	27	打开摄像头
KEYCODE_CALL	5	拨号
KEYCODE_ENDCALL	6	挂机
KEYCODE_FOCUS	80	拍照对焦键
KEYCODE_NOTIFICATION	83	通知键
KEYCODE_MUTE	91	使话筒静音

2. 控制键

控制键见表 6-3。

表 6-3 控制键

功 能 键	Keycode	描 述
KEYCODE_ENTER	66	回车键
KEYCODE_ESCAPE	111	Esc 键
KEYCODE_DPAD_CENTER	23	导航键，确定键
KEYCODE_DPAD_UP	19	导航键，向上
KEYCODE_DPAD_DOWN	20	导航键，向下
KEYCODE_DPAD_LEFT	21	导航键，向左
KEYCODE_DPAD_RIGHT	22	导航键，向右
KEYCODE_MOVE_HOME	122	用于把光标移动到开始
KEYCODE_MOVE_END	123	用于把光标移动到末尾
KEYCODE_PAGE_UP	92	向上翻页键
KEYCODE_PAGE_DOWN	93	向下翻页键
KEYCODE_DEL	67	退格键
KEYCODE_FORWARD_DEL	112	删除键
KEYCODE_INSERT	124	插入键

续表

功　能　键	Keycode	描　　述
KEYCODE_TAB	61	Tab 键
KEYCODE_NUM_LOCK	143	数码锁定键
KEYCODE_CAPS_LOCK	115	大写锁定键
KEYCODE_BREAK	121	Break/Pause 键
KEYCODE_SCROLL_LOCK	116	滚动锁定键
KEYCODE_ZOOM_IN	168	放大键
KEYCODE_ZOOM_OUT	169	缩小键
KEYCODE_BACKSPACE	67	退格键

3. 组合键

组合键见表 6-4。

表 6-4　组合键

功　能　键	描　　述
KEYCODE_ALT_LEFT	Alt+Left
KEYCODE_ALT_RIGHT	Alt+Right
KEYCODE_CTRL_LEFT	Ctrl+Left
KEYCODE_CTRL_RIGHT	Ctrl+Righ
KEYCODE_SHIFT_LEFT	Shift+Left
KEYCODE_SHIFT_RIGH	Shift+Right

4. 基本键

基本键见表 6-5。

表 6-5　基本键

功　能　键	Keycode	描　　述
KEYCODE_0	7	按键 0
KEYCODE_1	8	按键 1
KEYCODE_2	9	按键 2
KEYCODE_3	10	按键 3
KEYCODE_4	11	按键 4
KEYCODE_5	12	按键 5
KEYCODE_6	13	按键 6

续表

功 能 键	Keycode	描 述
KEYCODE_7	14	按键 7
KEYCODE_8	15	按键 8
KEYCODE_9	16	按键 9
KEYCODE_A	17	按键 A
KEYCODE_B	18	按键 B
KEYCODE_C	19	按键 C
KEYCODE_D	20	按键 D
KEYCODE_E	21	按键 E
KEYCODE_F	22	按键 F
KEYCODE_G	23	按键 G
KEYCODE_H	24	按键 H
KEYCODE_I	25	按键 I
KEYCODE_J	26	按键 J
KEYCODE_K	27	按键 K
KEYCODE_L	28	按键 L
KEYCODE_M	29	按键 M
KEYCODE_N	30	按键 N
KEYCODE_O	31	按键 O
KEYCODE_P	32	按键 P
KEYCODE_Q	33	按键 Q
KEYCODE_R	34	按键 R
KEYCODE_S	35	按键 S
KEYCODE_T	36	按键 T
KEYCODE_U	37	按键 U
KEYCODE_V	38	按键 V
KEYCODE_W	39	按键 W
KEYCODE_X	40	按键 X
KEYCODE_Y	41	按键 Y
KEYCODE_Z	42	按键 Z

6.2.3　Android 坐标点简介

1．坐标点基础知识

与 Web 系统不同，Android 系统中，屏幕的左上角是坐标系统的原点，坐标是（0,0）。原点向右延伸是 x 轴正方向，原点向下延伸是 y 轴正方向，如图 6-47 所示。

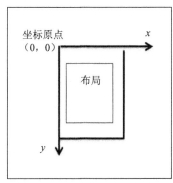

图 6-47　Android 坐标

2．坐标点的使用

首先，将手机调整到开发者选项，详见 6.1.1 节。

然后，打开"显示触摸操作"和"指针位置"。

触摸界面任意一点即可获得该触点坐标，并在屏幕上方显示具体数据，如图 6-48 所示，即可得到触点坐标 $X=169$，$Y=849$。

参数解释如下。

- P：按下屏幕。
- X：当手没松开时，为当前触点 x 坐标；当手松开时，为第一个点的相对 x 坐标（要求中间没松开）。
- Y：当手没松开时，为当前触点 y 坐标；当手松开时，为第一个点的相对 y 坐标（要求中间没松开）。

图 6-48　坐标点触摸

- Xv：沿 x 轴滑动的速度（变红表示手指操作是滑动，不变色表示单击）。
- Yv：沿 y 轴滑动的速度（变红表示手指操作是滑动，不变色表示单击）。

6.3 Android 自动化测试前的准备

前面讲述了坐标的相关知识,坐标是自动化测试中对象识别的重要手段。当然,这并不是唯一手段。接下来,我们要对 Android 系统上的对象、控件进行更为深入的研究。自动化测试可以通过除了坐标外的其他多种途径识别元素,如通过控件的类型(Class)、控件的 ID 来对控件进行识别。首先,需要明白页面的布局,熟悉各种组件,知道组件属性的意思。

6.3.1 布局

下面介绍 Android 中常用的布局。

(1)线性布局(LinearLayout)是将子组件按照垂直(Vertical)或者水平(Horizontal)两种方向来布局。可以在属性"orientation"中查看布局方向。另外一个常用的属性为"gravity",它用来控制上下左右的对齐方式,其属性值有上(Top)、下(Bottom)、左(Left)、右(Right)。

(2)帧布局(FrameLayout)是从屏幕的左上角(0,0)开始布局,多个组件层叠排序,后面的组件覆盖前面的组件。帧布局包括以下几种布局方式。

- 表格布局(TableLayout)。
- 相对布局(RelativeLayout)。
- 网格布局(GridLayout)。
- 绝对布局(AbsoluteLayout)。

6.3.2 Android 的组件

Android 的组件主要包括以下部分。

- 文本框 TextView。
- 编辑框 EditView。
- 按钮 Button。
- 单选按钮 RadioButton。

- 复选框 CheckBox。
- 状态开关按钮 ToggleButton。
- 开关 Switch。
- 拖动条 SeekBar。
- 时钟 AnalogClock 与 DigitalClock。
- 计时器 Chronometer。
- 列表视图 ListView。
- 网格视图 GridView。
- 进度条 ProgressBar。
- 星级评分条 RatingBar。
- 提示信息框 Toast。
- 滚动视图 ScrollView。

6.3.3 组件属性

Android 组件的属性见表 6-6。

表 6-6 Android 组件的属性

属性值	值类型	例子	属性值	值类型	例子
index	Int	0	enabled	Boolean	false
instance	Int	5	focusable	Boolean	false
class	String	android.widget.TextView	focused	Boolean	false
package	String	com.jike.test	Scrollable	Boolean	false
content-desc	String	string	Long-clickable	Boolean	false
checkable	Boolean	false	password	Boolean	false

6.3.4 确定包名和 Activity 值

Appium 测试需要提前配置 Package 和 Main Activity，而且在测试的执行过程中，经常涉及应用程序的切换和 Activity 的跳转。确定包名和 Activity 值的方式有很多种，这里针对只有源码、只有 APK 和 APK 已经安装完成 3 种情况进行介绍。本节实例使用的是中国移动的 APK，存放在笔者计算机的 E 盘上，绝对路径为"E:/ chinamobile.apk"。

1. 只有源码的情况

有源代码的应用程序，一般都是自己公司开发的 App，可非常容易地知道包名。直接打开 AndroidManifest.xml[①] 文件，找到"package="就可以知道包名，找到包含 android.intent.action.MAIN 和 android.intent.category.LAUNCHER 的 Activity。

2. 只有 APK 的情况

以网上下载的手机百度 APK（shoujibaidu_38274304.apk）为例，这里只有 APK 文件，笔者将下载的 APK 放置在 F 盘中，完整目录为 F:\shoujibaidu_38274304.apk。确定包含和 Activity 值的方式有以下两种。

方式一：采用 aapt 子命令查看包名和 Main Activity 名。

aapt 即 Android Asset Packaging Tool，位于 SDK 的 build-tools 目录下。如图 6-49 所示，其中 19.0.1 为下载的 Android API 版本。

图 6-49　aapt 所在目录

[①] Android 系统的清单文件，包含所有组件列表、应用需要申请的权限、应用安装对 Android 的要求等。在源码中，AndroidManifest.xml 位于应用工程的根目录中。

命令行格式为 aapt dump badging <file_path.apk>，其中 file_path.apk 为 APK 文件所在的绝对路径。

将命令行目录切换到本机 aapt 所在 android-sdk 的 build-tools 子目录下，如图 6-50 所示。

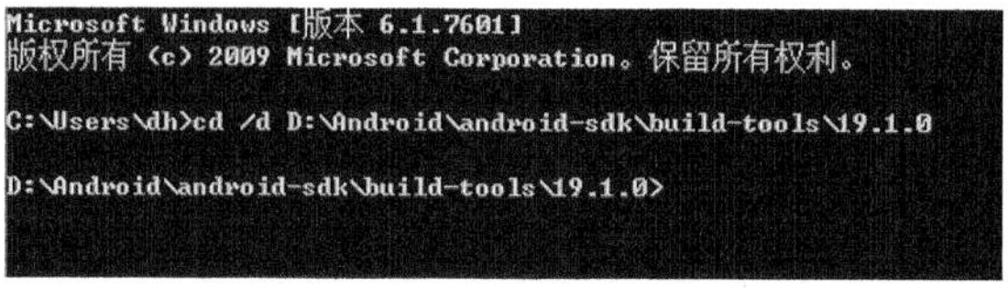

图 6-50　切换到 aapt 所在目录

输入如下命令。

```
aapt dump badging shoujibaidu_38274304.apk |findstr ①package
```

通过 `package:name='com.baidu.searchbox'` 获得待测试 APK 的包名，如图 6-51 所示。

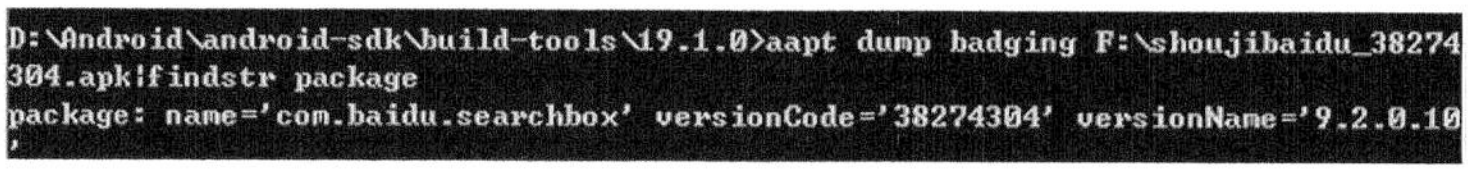

图 6-51　通过 aapt 获取包名

输入如下命令。

```
aapt dump badging shoujibaidu_38274304.apk |findstr activity
```

通过 `launchable-activity: name='com.baidu.searchbox.SplashActivity'` 获得 Main Activity，如图 6-52 所示。

图 6-52　通过 aapt 获取 Main Activity

 aapt 可以查看、创建、更新 ZIP 格式的文档附件(zip、jar、apk)。dump 表示查看 APK 包内指定的内容。badging 表示显示 APK 包中的 label 和 icon。

---

① "|" 管道命令，前一个命令的结果作为后一个命令的输入，findstr 用于在 aapt 子命令返回的结果中查找指定的字符串。

方式二：使用 Re-Signer 工具。

Re-Signer 是 APK 重签名[①]工具，也可以用于在 Android 自动化测试中获取包名和 Main Activity。

Re-Signer 下载地址如图 6-53 所示。

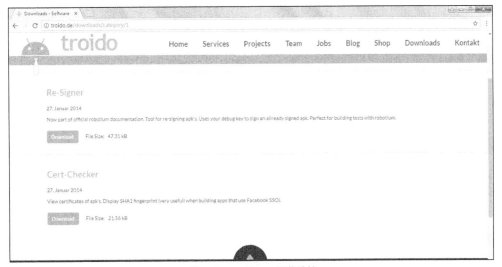

图 6-53　Re-Signer 下载地址

下载完成后，直接双击打开文件（re-sign.jar），如图 6-54 所示。

图 6-54　Re-Signer 工具打开页面

[①] Android 系统要求每一个 Android 应用程序必须要经过数字签名才能够安装到系统中，也就是说，如果一个 Android 应用程序没有经过数字签名，它是没有办法安装到系统中的。所有的 APK 都是有签名的，有些测试工具（如 Robotium）对 APK 进行测试时，需要重新签名。而 Appium 自身不需要 Debug 签名的限制，重签名的方法本书不做介绍。

将待测 APK shoujibaidu_38274304.apk 拖到 Re-Signer 工具打开页面指定位置后，Re-Signer 工具会重签名 APK，单击"保存"按钮保存重签名后的 APK。界面会提示包名和 Main Activity 信息，如图 6-55 所示。

图 6-55 通过 Re-Signer 获取包名和 Main Activity 名

3．APK 已经安装在手机上

在 APK 已经安装在手机上时，可以使用以下两种方式查看包名和 Activity 值。

方式一：使用 logcat。

（1）清除 logcat 内容，使用命令 adb logcat -c（日志信息量非常多，清除是为了更方便地找到有用信息）。

（2）启动 logcat，使用命令 adb logcat ActivityManager:I *:s。

（3）启动要查看的程序（手机百度），查看包名和 Main Activity 值，如图 6-56 所示。

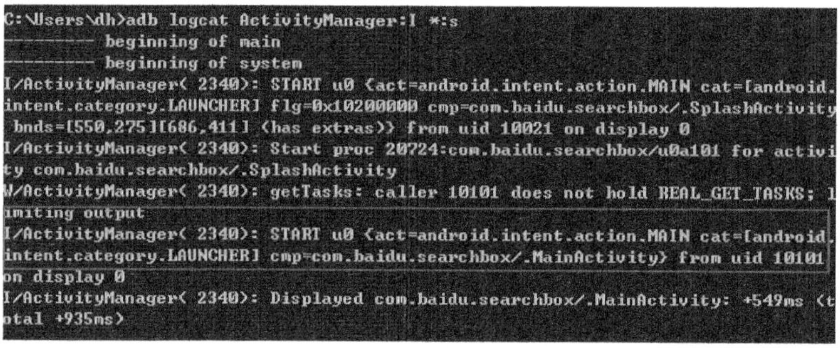

图 6-56 以 logcat 方式获取包名和 Main Activity 值

方式二：使用 dumpsys。

（1）启动要查看的程序：手机百度。

（2）在命令行输入以下内容。

```
adb shell dumpsys window w |findstr \/ |findstr name=
```

执行结果如图 6-57 所示。

图 6-57　通过 dumpsys 获取包名和 Main Activity 值

> Android 系统启动时会有大批的服务随之启动，因此可以用 dumpsys 命令来查看每个服务的运行情况。

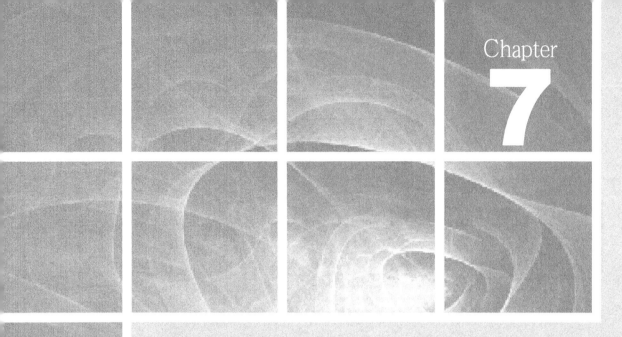

第 7 章

Android Appium 自动化框架

Appium 的设计哲学是这样的。

（1）不需要为了自动化而重新编译或者修改测试 App。

（2）不应该让移动端的自动化测试限制在某种语言和某个具体的框架下。任何人都可以使用自己最熟悉、最顺手的语言以及框架来做移动端的自动化测试。为了实现这个目的，Appium 选择了客户端/服务器的设计模式，只要客户端能够发送 HTTP 请求给服务器，那么客户端用什么语言实现都是可以的，通过这种方式 Appium 就可以支持多语言。

（3）不要为了移动端的自动化测试而重新发明轮子。对于 Web 端测试，WebDriver 协议中的 API 已经做到足够好了，Appium 没有必要重新实现一套协议，而是直接扩展了 WebDriver[①]协议，WebDriver 的 API 能够直接继承过来。

（4）移动端自动化测试工具应该是开源的。

7.1 Appium GUI 简介

Appium 是一个开源、跨平台的测试框架，可以用来测试原生项目、混合型项目以及移动端 Web 项目。Appium 支持 iOS、Android 平台。

备注	<ul><li>原生项目（Native App）：使用 iOS 或者 Android 平台开发的项目，如 QQ 手机客户端。</li><li>混合型项目（Hybrid App）：介于 Web App、原生 App 之间的 App，它虽然看上去是一个原生 App，但只有一个 UI Webview，里面访问的是一个 Web App，如百度客户端。</li><li>移动端 Web 项目（Mobile Web App）：需要使用移动端浏览器（Android 平台上支持 Chrome 和内置浏览器）打开的项目，如使用浏览器打开百度。</li></ul>

Appium 支持 Selenium WebDriver 支持的所有语言，如 Java、Object-C、JavaScript、PHP、Python、Ruby、C#、Clojure，以及 Perl 语言，还可以使用 Selenium WebDriver 的 API。Appium 实现了真正的多语言自动化测试。

7.2 Appium 架构详解

在 Android 上的架构，Appium 使用了 WebDriver 的 Json Wire[②]协议，来驱动 Android 系

---

① WebDriver：Selenium Webdriver 已经成为 Web 浏览器自动化测试的标准，也成了 W3C 的标准。
② Json Wire：客户端到服务器端的一种协议。

统的 UIAutomator[①]框架。Appium 也集成了 Selendroid，来支持老的 Android 版本，如图 7-1 所示。

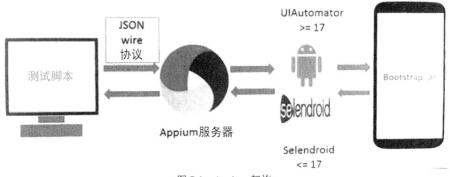

图 7-1　Appium 架构

Appium 是基于客户端/服务器架构的。

1）Appium 服务器

Appium 服务器的功能主要是监听一个端口，接收由客户端发送来的命令，并翻译这些命令。把这些命令翻译成移动设备可以理解的形式并发送给移动设备，然后由移动设备执行这些命令。执行命令后，把执行结果返回给 Appium 服务器，Appium 服务器再把执行结果返回给客户端。

2）Appium 客户端

Appium 客户端可以理解为发起命令的设备，一般来说，就是执行 Appium 测试代码的机器。狭义上，可以把客户端理解成代码，这些代码可以是 Java、Ruby、Python、JavaScript、PHP、C#等代码，只要它实现了 WebDriver 标准协议就可以。由于原生的 WebDriver API 是为 Web 端设计的，因此在移动端用起来会有点不伦不类。Appium 官方提供了一套 Appium 客户端，涵盖多种语言，如 Java、Ruby、Python、JavaScript、PHP、C#等。在测试的时候，一般要使用这些 Client 库去替换原生的 WebDriver 库。这实际上不是替换，仅是客户端对原生 WebDriver 进行了一些移动端的扩展，加入了一些方便的方法，如 Swipe（滑动操作）等。这样的设计思想带来了两个好处：带来多语言的支持；把服务器放在任意机器上，哪怕是云服务器上都可以。

[①] Uiautomator：准确来说，是 Google 在 Android 4.1 版本推出的一个测试的 Java 库，其中包含创建 UI 测试的各种 API 和自动化测试引擎，用来进行移动端 UI 自动化测试。

3）Session

Session 就是一个会话，Appium 中所有工作永远都是在 Session 启动后才可以进行的。一般来说，客户端初始化一个和服务器端交互的 Session，发送一个附有"Desired Capabilities"的 Json 对象参数的 POST 请求"/session"。

服务器端收到该数据后，会向客户端返回一个全局唯一的 Session ID。以后几乎所有的客户端请求都必须带上这个 Session ID，因为这个 Session ID 代表了你所打开的浏览器或者移动设备版本号等信息。

4）Desired Capabilities

Desired Capabilities 携带了一些配置信息，用来通知服务器建立需要的 Session。从本质上来讲，配置信息是 Key-Value（键值对）形式的对象。可以把它理解成是 Java 里的 Map，Python 里的字典，Ruby 里的 Hash，以及 JavaScript 里的 Json 对象。实际上，Desired Capabilities 在传输时就是 Json 对象。

Desired Capabilities 最重要的作用是告诉服务器本次测试的上下文，是要进行浏览器测试还是移动端测试。如果是移动端测试，指出是测试 Android 还是 iOS。如果测试 Android 指出要测试哪个 App。对于服务器的这些疑问，Desired Capabilities 都必须给予解答。

5）UIAutomator

Appium 在 Android 上是基于 UIAutomator 实现测试的代理程序（Bootstrap.jar）。当测试脚本运行时，每行指令都转换成 Appium 指令并发送给 Appium 服务器。然后，Appium 服务器将测试指令翻译后交给代理程序，将由代理程序负责执行测试。

6）Bootstrap.jar

Bootstrap.jar 是运行在 Android 手机上的一个应用程序，在手机上扮演 TCP 服务器的角色。当 Appium 服务器需要运行命令时，Appium 服务器会与 Bootstrap.jar 建立 TCP 通信，并把命令发送给 Bootstrap.jar。Bootstrap.jar 负责运行测试命令。

7.3 Appium Windows 环境搭建

在开始使用 Appium 实例之前，请先安装以下软件，见表 7-1。

表 7-1　Appium 环境软件

软件名称	是否安装	描述
Node.js	×	选装，需要以命令行形式运行 Appium 的时候才需要安装，参考 7.3.1 节
Android SDK	√	必装，参考 5.1 节
Eclipse	√	必装，方便使用 Java 语言编写测试代码，参考 3.2 节
TestNG	√	选装，也可使用 Junit，本书采用 TestNG 管理测试用例，参考 5.3 节
Appium	√	必装，参考 7.3.3 节

7.3.1　Node.js 的安装

Appium 是用 Node.js 编写的，若要用命令行启动 Appium 服务器，就必须安装该软件。下载地址为 Node.js 官网，如图 7-2 所示。笔者以 64 位 Windows 7 操作系统为例介绍 Appium 的运行环境。读者可以根据自己的测试环境进行选择性下载。

图 7-2　Node.js 安装包官方下载页面

安装过程中，一直单击 Next 按钮进行安装，如图 7-3 所示。

图 7-3　启动 Node.js 安装向导

勾选 I accept the terms in the License Agreement 复选框，并单击 Next 按钮，如图 7-4 所示。

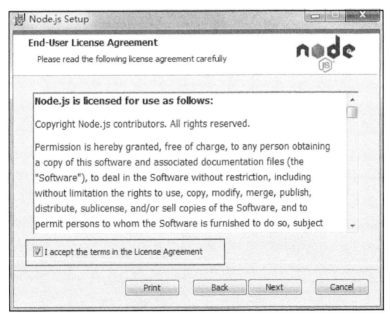

图 7-4　接受 Node.js 许可协议

输入 Node.js 安装目录，如图 7-5 所示。

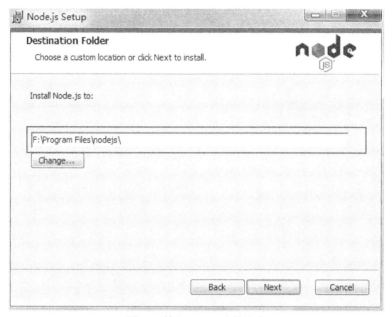

图 7-5　输入 Node.js 安装目录

单击 Next 按钮，选择要安装的 Node.js，如图 7-6 所示。

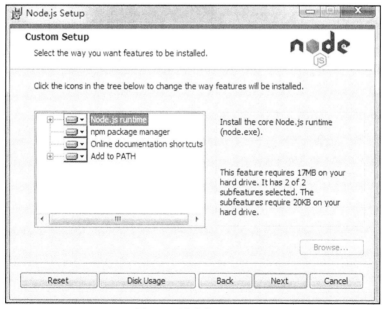

图 7-6　开始安装 Node.js

单击 Install 按钮进入安装步骤，如图 7-7 所示。

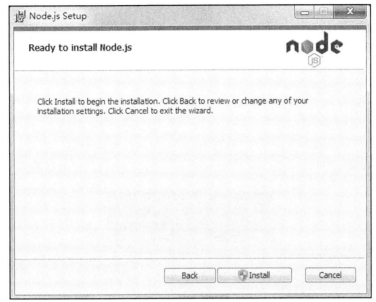

图 7-7　开始 Node.js 安装

安装完成，如图 7-8 所示。

图 7-8　Node.js 安装完成

Node.js 的安装不是必需的，但是若要用命令行启动 Appium 服务器就必须安装它。

下面测试安装是否成功。

在安装过程中，安装程序会自动配置环境变量，无须手工配置。安装完成后，打开 Windows 命令行窗口，通过执行命令 node -v，验证是否可以进入交互模式。如果添加成功会进入交互模式，如图 7-9 所示。

图 7-9　验证 Node.js 安装是否成功

若返回版本号信息，则说明 Node.js 安装成功。

7.3.2　.NET Framework 的安装

安装 Appium 的 Window 版本会提示安装.NET Framework 4.5，单击"是"按钮直接下载并安装，或者从微软官网下载并安装，如图 7-10 所示。

图 7-10　选中.NET Framwork 4.5 下载

下载完成后，双击开始安装软件，如图 7-11 所示。

图 7-11 安装.NET Framework

单击"安装"按钮进行软件的安装,如图 7-12 所示。

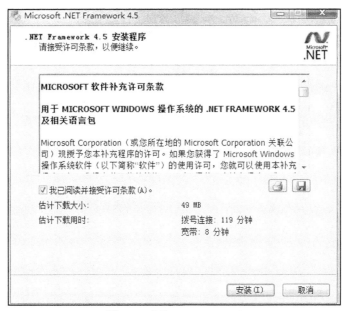

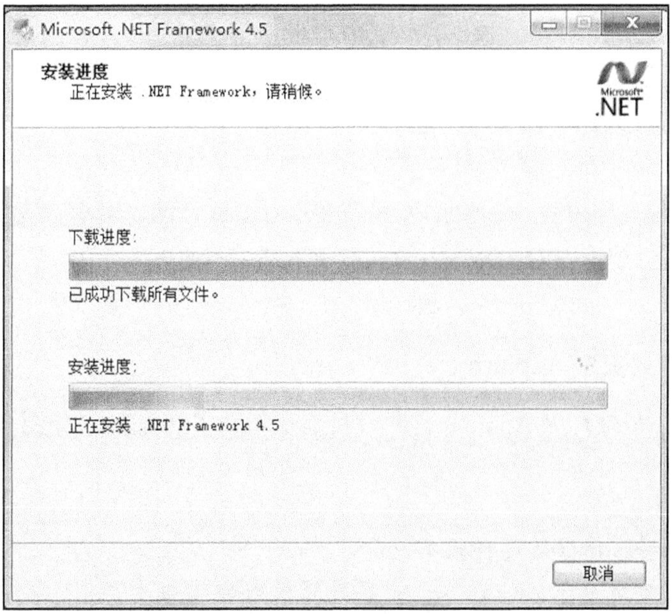

图 7-12 查看.NET Framework 安装进度

安装完成后的界面,如图 7-13 所示。

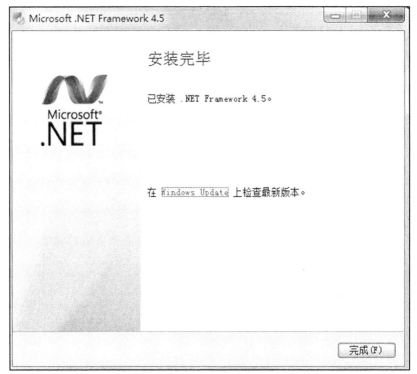

图 7-13　安装.NET Framework 完成

7.3.3　Appium 的安装与配置

Appium 有两种安装方式——GUI 方式和命令行方式，两者没有太大的区别。GUI 格式带有 GUI，可比较直观地查看；命令行格式方便在自动化测试过程中启动和关闭 Appium 服务器。

1．以 GUI 方式安装 Appium

Appium GUI 下载地址为 Appium 官网，目前最新版本为 1.2.3，如图 7-14 所示。

下载时，请注意选择版本。
这里下载 Windows 版本，如果计算机为 Mac 版本，则下载 dmg 格式。

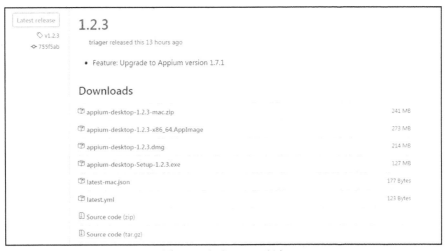

图 7-14 查看 Appium 版本

下载完成后，解压到任意位置，双击 appium-desktop-Setup-1.2.3.exe 自动安装，如图 7-15 所示。

图 7-15 Appium 安装文件

安装完成后，将 Appium 的路径和 appium-doctor 路径放置到环境变量中。例如，这里安装路径是 F 盘下面。

在环境变量 Path 中增加"；F:\Appium;F:\Appium\node_modules\.bin;"。环境变量配置成功后，在命令行运行"appium-doctor"。如果弹出图 7-16 所示界面，则说明 Appium 安装成功。

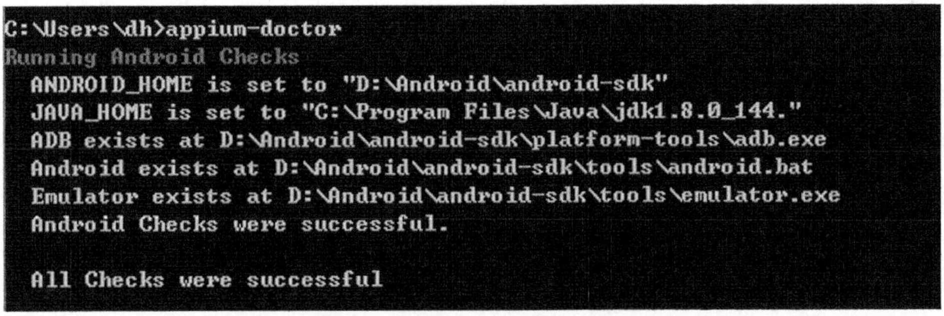

```
C:\Users\dh>appium-doctor
Running Android Checks
ANDROID_HOME is set to "D:\Android\android-sdk"
JAVA_HOME is set to "C:\Program Files\Java\jdk1.8.0_144."
ADB exists at D:\Android\android-sdk\platform-tools\adb.exe
Android exists at D:\Android\android-sdk\tools\android.bat
Emulator exists at D:\Android\android-sdk\tools\emulator.exe
Android Checks were successful.

All Checks were successful
```

图 7-16 验证 Appium 安装是否成功

安装成功后，双击打开 Appium，查看 Appium 界面，如图 7-17 所示。

图 7-17　Appium 界面

2．以命令行方式安装 Appium

安装好 Node.js 后,可以采用命令行的方式安装 Appium。在 Windows 命令行工具中输入如下命令。

```
npm install -g appium
```

因为 npm 的服务器在国外,所以下载和安装速度较慢。可以采用淘宝源进行安装,命令如下。

```
npm install appium -g --registry=https://registry.npm.taobao.org
```

要安装具体的版本(Appium 版本为 1.6.3),命令如下。

```
npm install -g appium@1.6.3
```

等待安装完毕,在某些情况下,安装 Appium 的时候并不会把 Appium 的路径放进系统的 Path 里,这时候需要手工添加一下。在环境变量的 Path 变量名下,增加 Appium 的安装路径,如图 7-18 所示。

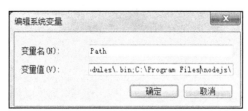

图 7-18　Appium 环境变量配置

和以 GUI 方式安装 Appium 一样,安装后,为了验证 Appium 是否安装成功,可在 Windows

命令行工具中输入命令"appium-doctor"。

7.4 Appium GUI 详解

1. 通过 Simple 标签启动 Appium 服务器

打开 Appium 界面后，可以看到服务器启动窗口。如果服务器端地址为本机，则直接单击 Start Server v1.7.0 按钮启动 Appium 服务器，也可以根据需要修改服务器端主机地址和端口号。笔者的服务器端直接运行在本机上，使用默认的 Host 和 Port，如图 7-19 所示。

图 7-19　Appium 界面

2. 通过 Advanced 标签启动 Appium 服务器

在 Appium 界面中，单击 Advanced 标签，设置 Appium 中可用的所有服务器标志，如图 7-20 所示。

图 7-20　Advanced 选项卡

使用 Advanced 选项完成配置后，单击 Start Server v1.7.0 按钮启动 Appium 服务器，单击 Save As Preset 按钮可以保存预设的 Advanced 选项，方便以后使用。已经保存的预设可以在 Presets 选项卡中看到，也可以在 Presets 选项卡直接启动 Appium 服务器。

3．通过 Presets 标签启动服务器

如果使用了配置项 Advanced，就可以保存配置以备以后使用，已经保存在 Presets 的预设列表中，如图 7-21 所示。在以后的使用中可以根据需要启动不同的预设，而无须重新配置。选中一条预设记录，单击 Start Server v1.7.0 按钮启动 Appium 服务器，单击 Delete Preset 按钮删除一条预设记录。

启动服务器后，Appium 服务器将在指定的 IP 和端口上运行，Appium GUI 如图 7-22 所示，用于显示服务器日志。

启动服务器后，用户可以根据需要直接复制有用的日志到测试报告中（直接在有用日志上，按<Ctrl+C>快捷键进行复制）。运行后的界面控件如下。

图 7-21　Presets 选项卡

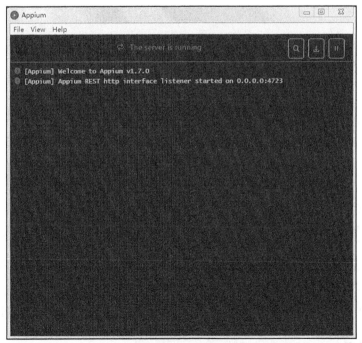

图 7-22　Appium GUI

Start Session：启动新的会话窗口，在当前运行的服务器上，启动 Inspector 查找控件会话，如图 7-23 所示。

除此之外，在 Appium 界面中，选择 File→New Session Window，可以不必启动本地服务器，直接打开新的会话窗口，如图 7-23 所示。

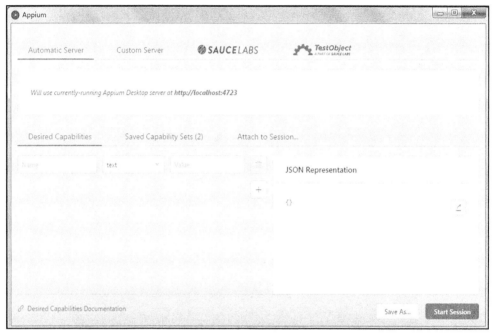

图 7-23　新会话窗口

7.5　新会话窗口

1．Appium 端点

针对非本地运行的 Appium 服务器启动会话，需要进行如下设置。

（1）Custom Server。自定义主机 IP（Remote Host）/端口（Remote Port）设置，如果想在本机启动 Inspector 查找控件元素，而 Appium 服务器运行在网络中其他设备上，这个设置是非常有用的，如图 7-24 所示。

（2）SAUCELABS[①]。如果没有测试设备（如 Android 模拟器），则可以通过 Sauce 实验

---

① Sauce Labs：一个提供自动化功能测试的云测试服务公司。

室账号登录云服务来创建一个 Appium 会话，如图 7-25 所示。

图 7-24　Custom Server 选项卡

图 7-25　SAUCELABS 选项卡

（3）TestObject[①]。利用 TestObject 真机云服务可以创建基于真实设备的 Inspector 会话，如图 7-26 所示。

图 7-26　TestObject 选项卡

2. Desired Capabilities

Desired Capabilities 是启动 Session 的时候必须提供的，它告诉 Appium 服务器端自动测试平台是什么、App 存放位置等。Desired Capabilities 本质上是键值对，以 JSON 的格式发送到 Appium 服务器端。具体的字段解释，参考 7.8.1 节。

1）Desired Capabilities

Appium GUI 提供了友好的 Desired Capabilities 设置方式并保持它们供以后使用。在

---

① TestObject：TestObject 服务由柏林一家公司创立，所提供的云服务能够让应用开发商在一系列 Android 设备上自动远程测试其应用。

Desired Capabilities 选项卡下面以表单的形式进行设置，如图 7-27 所示，根据需要单击 "+" 按钮，增加 Desired Capabilities 选项。JSON Representation 文本框中显示已设置 Desired Capabilities 项的 JSON 格式，这是实际发送给 Appium 服务器的内容。

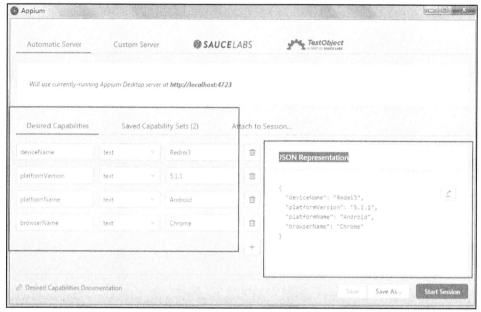

图 7-27　Desired Capabilities 设置

Desired Capabilities 设置完成后，如果希望保存它们，以便以后运行相同类型的会话，单击 Save As 按钮，在弹出的对话框中设定待保存的 Desired Capabilities 名字。单击 Save 按钮进行保存，如图 7-28 所示。已保存的配置可以在 Saved Capabilities Sets 选项卡下访问，便于以后编辑或启动会话。

图 7-28　输入 Desired Capabilities 名称

2）Saved Capabilities Sets

已经保存的 Desired Capabilities 设置，可以在该选项卡下查看，如图 7-29 所示。左侧显示 Desired Capabilities 设置项列表，选中其中一个记录，右侧会显示已设置 Desired Capabilities

项的 JSON 格式。单击记录后的 ✎ 图标，页面跳转到 Desired Capabilities 选项卡以修改配置。单击记录后的 🗑，删除该条记录。

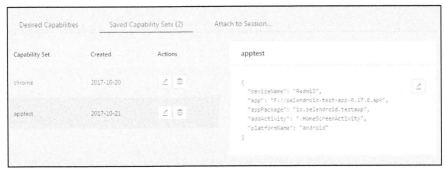

图 7-29　Saved Capability Sets 选项卡

3）Attach to Session

当前正在运行的所有会话都会显示在 Attach to Session 选项卡下，通过选中会话 ID 的方式选中不同的会话进行启动。如果在测试运行过程中调试这个功能将非常有用，虽然 Inspector 窗口已经关闭，但是保存在服务器端的会话不会关闭。可以通过这种方式再次打开会话，查看页面元素，如图 7-30 所示。

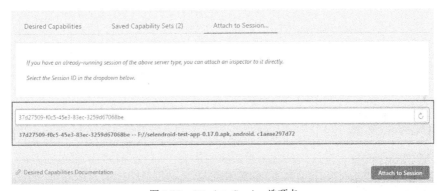

图 7-30　Attach to Session 选项卡

7.6　在 Appium 中查找控件

有很多读者可能会问如何获取手机应用活动各种组件的信息。最常见的是使用 UI Automator Viewer[①]，本节将重点介绍使用 Appium 自带的 Inspector 工具的应用。

---

① UI Automator Viewer：一个用来扫描和分析 Android 应用程序的 UI 组件的 GUI 工具。

7.6.1 Appium Inspector 界面

在做自动化测试时，经常在代码执行过程中因为选择的元素定位方式不准确而无法唯一地定位元素，导致测试执行报错。于是，需要修改定位方式后再次运行，查看运行结果，这样会导致时间的浪费。如何在测试执行前查看我们选择的元素定位策略能否唯一地查找到控件呢？Appium 集成了查找控件的办法帮我们解决这个问题。例如，使用 selendroid-test-app-0.17.0.apk 定位 email 控件。

在图 7-27 中设置 Desired Capabilities，设置示例见表 7-2。

表 7-2　Desired Capabilities 设置示例

Desired Capabilities 项	Desired Capabilities 值	含　义
deviceName	Redmi3	待测设备名为 Redmi3
app	F://selendroid-test-app-0.17.0.apk	待测 App 所在目录
appPackage	io.selendroid.testapp	待测 App 的包名
addActivity	.HomeScreenActivity	待测 App 的启动 Activity
platformName	android	待测设备平台

设置完成后单击 Start Session 按钮，启动一个会话，如图 7-31 所示。

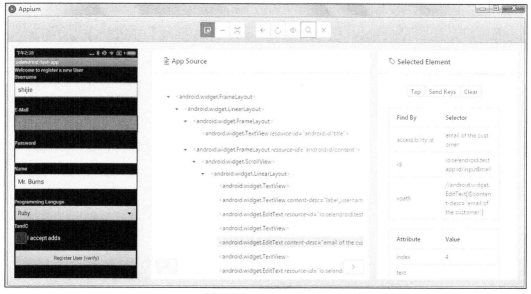

图 7-31　Appium Inspector

加载 Appium Session 可能需要一些时间，尤其是云服务。当 Session 加载完成后，App 的截图会出现在应用程序的左边。可以用鼠标选择界面元素，被选中的元素将高亮显示，如图 7-31 所示。

中间的 App Source 面板显示元素的层级结构，以 XML 方式展示，可以在这个层级结构上选中元素，左边的界面上同样会高亮显示。

当一个元素高亮显示时，右边的 Selected Element 面板显示的是选中元素的细节，这个细节包括识别方式、操作按钮和属性列表。

7.6.2 Selected Element 面板

当元素高亮显示时，还可以对元素进行操作，如单击 Tap 按钮完成单击操作，单击 Send Keys 按钮完成发送数据（如果它是一个文本控件）。例如使用 selendroid-test-app-0.17.0.apk 定位 email 控件。

1．操作按钮

操作按钮如图 7-32 所示。多个按钮的功能如下。

- Tap 按钮：单击被选元素。
- Send Keys 按钮：对被选元素赋值（针对文本控件）。
- Clear 按钮：清空备选元素值。

当对元素进行操作时，Inspector 会发送一系列的命令给 Appium，Appium 将执行这些操作。如果执行成功，更新截图；如果失败，弹出报错信息。例如单击"E-mail"中输入"sh***@126.com"值，执行如下步骤。

（1）单击 Send Keys 按钮。

（2）在弹出的 Send Keys 对话框中输入"sh***@126.com"，如图 7-33 所示。

图 7-32　操作按钮

图 7-33　Send Keys 对话框

（3）单击 Send Keys 按钮。如图 7-34 所示，更新截图界面，E-mail 字段赋值完成。

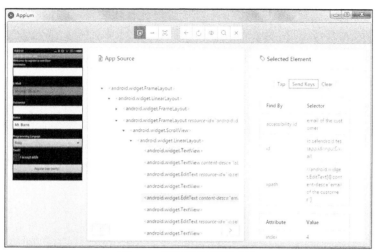

图 7-34　操作后页面截图更新

Appium 日志中也可以看到执行的相关信息，如图 7-35 所示。

图 7-35　Appium 日志

2．元素识别

元素识别一直是 UI 自动化测试的难点所在，Appium 在该表格中给出唯一定位元素的方式，方便书写自动化测试用例，如图 7-36 所示。值得注意的是，有时 Selector 列给出的值太过复杂，建议修改和确认后再使用。

3．属性值

该区域中显示元素的详细信息，如图 7-37 所示，为后面章节中讲述的元素定位方式提供依据。

图 7-36　元素识别　　　　　　　　图 7-37　元素属性值

7.6.3　操作区域

为了方便调试，Appium 的 Inspector 集成了以下操作，如图 7-38 所示。各个图标的功能如下。

图 7-38　操作区域的图标

- Select Elements：选中元素。选中后，当在截图上选中元素时，元素会高亮显示，同时中间层级结构高亮显示被选中元素对应的层级。右侧显示元素的详细信息。
- Swipe By Coordinates：通过坐标点滑动。选中后，截图上光标呈现"＋"形，同时在左上角显示光标所在位置的坐标。滑动操作后，Inspector 会发送一系列的命令给 Appium，Appium 将执行滑动操作。如果执行成功，更新截图；如果失败，则弹出报错信息。

- Tap By Coordinates：通过坐标点进行单击。选中后，截图上光标呈现"＋"形，同时在左上角显示光标所在位置的坐标。单击操作后，Inspector 会发送一系列的命令给 Appium，Appium 将执行滑动操作。如果执行成功，更新截图；如果失败，则弹出报错信息。

- Back：相对于调用 driver.back 方法，执行后退操作。

- Refresh：更新页面截图。

- Start Recording：启动录制。

- Search for element：查找元素。支持在 Android 上采用 id、xpath、name、classname、accessibility id、uiautomator selector 对元素进行查找。

- Quit the session：相当于调用 driver.quit，执行关闭 Inspector 的操作。

7.6.4 调试定位方式

在 Appium 的 Inspector 上，单击 Search for element 图标，弹出 Search for element 对话框，如图 7-39 所示。

图 7-39　Search for element 对话框

选择 Locator Strategy（定位方法）、Selector（定位值），单击 Search 按钮进行查找。

1. 未找到元素

查找完成后，标题 Elements 后的数据为查找到的数量。如果查找不到，编辑框中将提示

"Could not find any elements",并且标题 Elements 后的数据为 0,如图 7-40 所示。

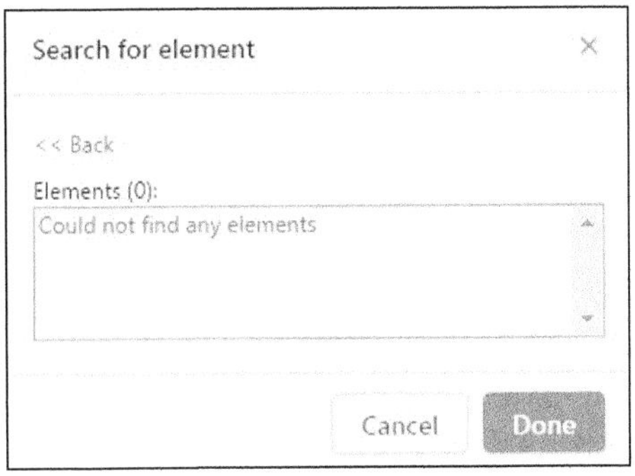

图 7-40　未查找到控件元素

2．查找到多个元素

找到多个元素后,标题 Elements 后显示找到的元素数量。编辑框中以列表的形式显示所有通过该定位方式查找到的元素记录,如图 7-41 所示。

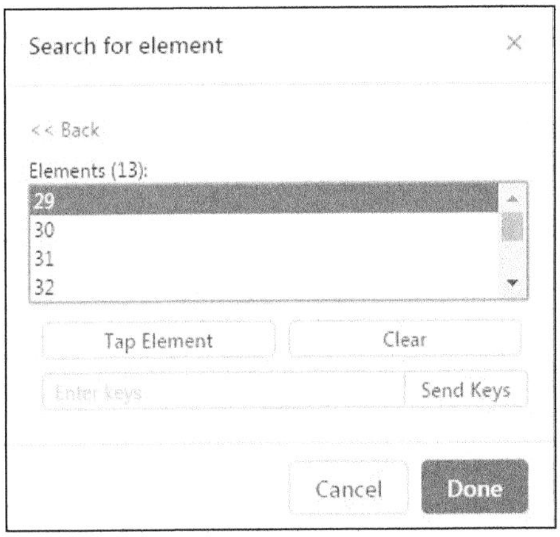

图 7-41　找到多个控件元素

在图 7-41 中,Elements(13) 中的 13 表示本次共查到了 13 个元素。下拉框的数字编号分别表示这 13 个元素在 Search for element 方法里的编号。根据笔者的经验,这个编号每次并不是都从 1 或者 0 开始顺次累加,起始编号也许从某一个数值开始。这次查询结果中第一个元素的编号就从 29 开始,总共有 13 个元素。

3．找到唯一的元素

如图 7-42 所示,Elements 后的元素数量为 1,编辑框中只有一条记录,表示通过该定位方式可以唯一定位到的元素。

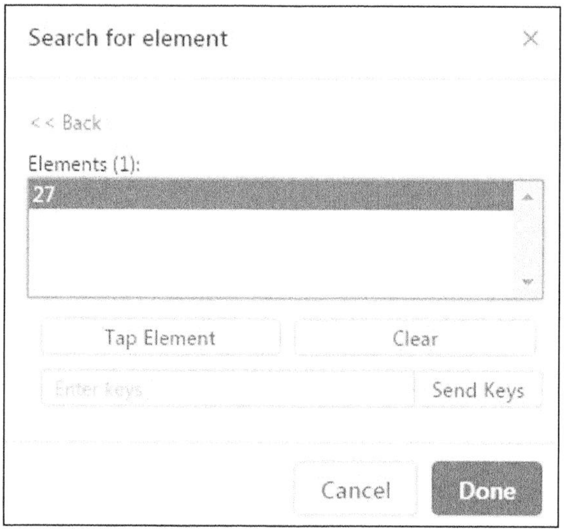

图 7-42　找到唯一的控件元素

4．操作元素

通过在编辑框中选择一条记录，操作按钮（Tap Element、Clear 和 Send Keys）会从灰色状态变成可用状态。可以对选中元素进行操作，如图 7-43 所示。

图 7-43　选中查找到的控件元素

在图 7-43 中，E-mail 控件为文本控件，可以进行输入操作，输入"sh***@126.com"后，单击 Send Keys 按钮。Appium 将执行这些操作命令。如果执行成功，则更新截图为操作命令执行后的截图，如图 7-44 所示；如果失败，将会弹出报错信息。

图 7-44　操作查找到的控件元素

7.7　Appium 录制功能

Appium 和其他自动化框架一样提供了非常基本的录制功能，这些录制的代码可以直接复制和粘贴到客户端，便于测试使用。通过录制可以用最简单的方式将人的手工动作转化为程序代码。在真正的自动化过程中，录制下来的自动化代码还需要修正，并不能直接使用，因为其成功率比较低。

在 Appium 界面中单击录制按钮开始进行录制。操作步骤为进入注册页面，在 Username 文本框中输入"sh***"，在 E-mail 文本框中输入"sh***@126.com"。

在 App Source 中选中元素，在 Selected Element 中进行各种操作，如输入值等。操作步骤会记录到 Recorder 中。执行结果如图 7-45 所示。

图 7-45　在 Appium 中使用录制功能

在 Recorder 后面的下拉框中可以选中对应的编程语言和框架。在 Java 环境下，系统默认选中 java-Junit 模式进行录制。

Appium 界面中 Show/Hide Boierplate Code 的作用是显示/隐藏模板代码。如果显示，代码中将会包含 Desired Capabilities 设置、Session 启动以及 Junit 模板代码；如果隐藏，只显示操作命令，如代码清单 7-1 所示。

代码清单 7-1　　　　　　　　　　隐藏模板代码

```
(new TouchAction(driver)).tap(552, 273).perform()
    MobileElement el1 = (MobileElement) driver.findElementById ("io.selendroid.testapp:id/
inputUsername");
    el1.sendKeys("shijie");
    MobileElement el2 = (MobileElement) driver.findElementById("io.selendroid.testapp:id/
inputEmail");
    el2.sendKeys("shijie@126.com");
  }
```

Appium 界面中 Copy Code to Chipboard 的作用是复制代码到剪切板上。

录制完成后，在 Java 项目目录下创建一个与录制代码中名字相同的类，并将录制的代码全部复制到已创建的类中。复制后的录制代码如代码清单 7-2 所示。

代码清单 7-2   录制脚本

```java
import io.appium.java_client.MobileElement;
import io.appium.java_client.android.AndroidDriver;
import junit.framework.TestCase;
import org.junit.After;
import org.junit.Before;
import org.junit.Test;
import java.net.MalformedURLException;
import java.net.URL;
import org.openqa.selenium.remote.DesiredCapabilities;

public class SampleTest {

  private AndroidDriver driver;

  @Before
  public void setUp() throws MalformedURLException {
    DesiredCapabilities desiredCapabilities = new DesiredCapabilities();
    desiredCapabilities.setCapability("platformName", "Android");
    desiredCapabilities.setCapability("platformVersion", "5.1.1");
    desiredCapabilities.setCapability("deviceName", "Redmi 3");
    desiredCapabilities.setCapability("appPackage", "io.selendroid.testapp");
    desiredCapabilities.setCapability("appActivity", ".HomeScreenActivity");
    desiredCapabilities.setCapability("app", "F:\\book\\selendroid-test-app-0.17.0.apk");

    URL remoteUrl = new URL("http://localhost:4723/wd/hub");

    driver = new AndroidDriver(remoteUrl, desiredCapabilities);
  }

  @Test
  public void sampleTest() {
```

```
    (new TouchAction(driver)).tap(552, 273).perform()
    MobileElement el1 = (MobileElement) driver.findElementById("io.selendroid.testapp:id/
    inputUsername");
    el1.sendKeys("shijie");
    MobileElement el2 = (MobileElement) driver.findElementById("io.selendroid.testapp:id/
    inputEmail");
    el2.sendKeys("shijie@126.com");
  }

  @After
  public void tearDown() {
    driver.quit();
  }
}
```

通过录制，用户可以更快地创建代码，并且给自动化创建者提供了将动作代码化的便捷方式。需要注意的是，录制的代码易用性不是很强，经过修改才可以使用，如"(new TouchAction(driver)).tap(552, 273).perform()"就是通过对坐标点的操作来完成的。若手机端的分辨率或者屏幕尺寸发生变化，该代码将不能使用。

7.8 Desired Capabilities 的配置

7.5 节已经介绍了 Desired Capabilities 的作用，本节将主要介绍 Desired Capabilities 如何配置。

7.8.1 Desired Capabilities 配置简介

Desired Capabilitie 简单来说就是用键值对的形式来实现用测试脚本控制 Appium 的行为。下面的一些通用配置是需要指定的。

与 Appium 服务相关的关键字，见表 7-3。

对于自动化引擎，很多初学者可能不太明白。Appium 中的 automationName 字段是用来设置自动化测试引擎的。

目前提供了 3 种自动化引擎。当 Android 版本小于 4.2 时，使用 SELENDROID 自动化引擎。4.2 版本以上默认的自动化引擎为 Appium。对于较新版本的 Android，建议采用 ANDROID_UIAUTOMATOR2 作为自动化引擎。

表 7-3 与 Appium 服务相关的关键字

关 键 字	描 述
automationName	使用的自动化测试引擎
platformName	测试的手机操作系统
platformVersion	手机操作系统版本
deviceName	使用的手机类型或模拟器类型
app	.ipa 或.apk 文件所在的本地绝对路径或者远程路径，也可以是一个包括两者之一的.zip 文件。Appium 会先尝试安装路径对应的应用在适当的真机或模拟器上。针对 Android 系统，如果指定 app-package 和 app-activity（具体见下面），那么就可以不指定 App，否则会与 browserName 冲突
browserName	需要进行自动化测试的手机 Web 浏览器名称。如果对应用进行自动化测试，这个关键字的值应为空
newCommandTimeout	设置命令超时时间，单位为秒。若达到超时时间时仍未接收到新的命令，Appium 会假设客户端退出，然后自动结束会话
language	设定模拟器的语言
locale	设定模拟器的区域设置
udid	连接的物理设备的唯一设备标识
orientation	在一个设定的方向模式中开始测试
autoWebview	直接转换到 Webview 上下文，默认值为 False
noReset	不要在会话前重置应用状态，默认值为 False
fullReset	对于 iOS，删除整个模拟器目录。对于 Android，通过卸载（而不是清空数据）来重置应用状态。在 Android 上，这也会在会话结束后自动清除被测应用。默认值为 False
eventTimeings	启动或者禁用 Appium 内部时序报告（如每个命令的启动和关闭等），默认值为 False，如果启用，设置为 True
enablePerformanceLogging	启用 chromeDriver 或者 Safari 的性能日志，默认值为 False，只适用于 Web 或者 Webview

仅对 Android 测试有效的关键字，见表 7-4。

表 7-4 仅对 Android 测试有效的关键字

关 键 字	描 述
appActivity	从应用包中启动的 Android Activity 名称。它通常需要在前面添加（如使用.MainActivity 而不是 MainActivity）
appPackage	运行的 Android 应用的包名
appWaitActivity	等待启动的 Android Activity 名称
deviceReadyTimeout	设置等待一个模拟器或真机准备就绪的超时时间
androidCoverage	用于执行测试的 instrumentation 类。作为命令 adb shell am instrument -e coverage true -w 的 -w 参数
androidCoverageEndIntent	用于导出覆盖率到文件，作为 adb shell am broadcast –a 的参数
androidDeviceReadyTimeout	（仅适用于 Chrome 和 webview）开启 Chromedriver 的性能日志（默认值为 False）
androidInstallTimeOut	等待 APK 安装成功的超时时间，默认为 90000，单位为毫秒
androidInstallPath	APK 在安装前，安装程序放置路径，默认为/data/local/tmp
adbPort	adb 服务器连接的端口
remoteAdbHost	远程 adb 服务器的地址
androidDeviceSocket	Devtools Socket 名称，仅用于被测 App 是内置浏览器的情况下
avd	需要启动的 AVD (安卓虚拟设备) 名称
avdLaunchTimeout	以毫秒为单位，等待 AVD 启动并连接到 ADB 的超时时间（默认值为 120000）
avdReadyTimeout	以毫秒为单位，等待 AVD 完成启动动画的超时时间（默认值为 120000）
avdArgs	启动 AVD 时需要加入的额外的参数
useKeystore	使用一个自定义的 keystore 来对 APK 进行重签名。默认值为 False
keystorePath	自定义 keystore 的路径。默认值为~/.android/debug.keystore
keystorePassword	自定义 keystore 的密码
keyAlias	key 的别名
keyPassword	key 的密码
chromedriverExecutable	WebDriver 可执行文件的绝对路径（如果 Chromium 核心提供了对应的 WebDriver，应该用它代替 Appium 自带的 WebDriver）
autoWebviewTimeout	以毫秒为单位，等待 Webview 上下文激活的时间。默认值为 2000
intentAction	用于启动 activity 的 intent action（默认值为 android.intent.action.MAIN）
intentCategory	用于启动 activity 的 intent category（默认值为 android.intent.category.LAUNCHER）
intentFlags	用于启动 activity 的标记（默认值为 0x10200000）

续表

关 键 字	描 述
optionalIntentArguments	用于启动 activity 的额外 intent 参数。请参考 Intent 参数
unicodeKeyboard	使用 Unicode 输入法。默认值为 False
resetKeyboard	在设定了 unicodeKeyboard 关键字的 Unicode 测试结束后，重置输入法到原有状态。如果单独使用，将会被忽略。默认值为 False
noSign	跳过检查和对应用进行 Debug 签名的步骤。只能在使用 UiAutomator 时使用，使用 Selendroid 时不行。默认值为 False
ignoreUnimportantViews	调用 uiautomator 的 setCompressedLayoutHierarchy 函数。由于 Accessibility 命令在忽略部分元素的情况下执行速度会加快，因此这个关键字能加快测试执行的速度。被忽略的元素将无法找到，因此这个关键字同时也被实现成可以随时改变的设置（Settings）。默认值为 False
disableAndroidWatchers	禁止监听应用程序不响应或者崩溃。禁止后会降低真机或者模拟器的 CPU 消耗，只适用于 uiAutomator，不适用于 selendroid
chromeOptions	允许传递 chromeOption 参数到 ChromeDriver
recreateChromeDriverSessions	当切换到非 ChromeDriver 环境时，关闭 ChromeDriver 会话
nativeWebScreenshot	在 Web 上下文中，采用本地（adb）截图，而不是采用代理。默认值为 False
androidScreenshotPath	屏幕截图保存路径，默认为/data/local/tmp
autoGrantPermissions	允许 Appium 自动获取 App 安装的所有权限，默认值为 False
networkSpeed	设置为网络带宽，是指最大上传和下载速度。默认值为 Full
gpsEnabled	在启动 Session 前，切换 GPS 定位，默认值为 Enable
isHeadless	当使用 Headless 模式时，设置为 True，默认值为 False。支持 iOS

以上基本包括现有 Capabilities 中的绝大多数内容，本节主要以 Android 为主进行测试，还有一部分与 iOS 相关的 Capabilities 在这里就不再详述了。

7.8.2 Desired Capabilities 配置示例

要配置 Desired Capabilities，可以通过代码初始化 Desired Capabilities 对象。本节以配置 selendroid-test-app-0.17.0.apk 为例，介绍 Desired Capabilities 的配置。

1. 原生 App 和混合 App 的 Desired Capabilities 配置示例

在配置 Desired Capabilities 之前，首先导入以下包。

```java
import java.io.File;
import org.openqa.selenium.remote.DesiredCapabilities;
import io.appium.java_client.remote.AndroidMobileCapabilityType;
```

下面介绍如何测试原生 App 和混合 App 的 Desired Capabilities 配置，如代码清单 7-3 所示。

代码清单 7-3　　　　原生 App 和混合 App 的 Desired Capabilities 配置

```java
DesiredCapabilities cap = new DesiredCapabilities();
//待测试的手机操作系统（Android, IOS）,这里设置为 "Android"
capabilities.setCapability ("platformName", "Android");
//手机操作系统版本
capabilities.setCapability ("platformVersion", "4.3");
//待使用的自动化测试引擎：Appium(默认)或 Selendroid
capabilities.setCapability ("automationName", "appium");
//使用的手机类型或模拟器类型,真机时输入 Android Emulator 或者手机型号（这里设置为"Lenovo A788t"）
capabilities.setCapability("deviceName", "Lenovo A788t");
//连接的物理设备的唯一设备标识,Android 可以不设置,这里设置为 "00a10399"
capabilities.setCapability ("udid", "00a10399" );
//设置收到下一条命令的超时时间,如果超时,Appium 会自动关闭 session,默认值为 60 秒
capabilities.setCapability ("newCommandTimeout", "300");
//支持中文输入,会自动安装 Unicode 输入法。默认值为 False
capabilities.setCapability ("unicodeKeyboard", "True");
//在设定了 unicodeKeyboard 关键字的 Unicode 测试结束后,重置输入法到原有状态
capabilities.setCapability ("resetKeyboard", "True");
//未安装应用时,设置 App 的路径
//若手机已安装 App,直接从手机启动 App,下面的路径不设置
capabilities.setCapability ("app", "D:\\ selendroid-test-app-0.15.0.apk");
//待运行的 Android 应用的包名
capabilities.setCapability("appPackage", "io.selendroid.testapp");
//待启动的 Android 应用对应的 Activity 名称,如 MainActivity 和.Settings
capabilities.setCapability("appActivity", ".HomeScreenActivity");
//等待启动的 Android Activity 名称,如 SplashActivity
capabilities.setCapability ("appWaitActivity", ".HomeScreenActivity");
driver = new AndroidDriver(new URL("http://127.0.0.1:4723/wd/hub"),capabilities);
```

2. Web App Capabilities 配置实例

在 Web App 中，有些在原生 App 或者混合 App 中的配置选项是用不到的，如 app、appPackage 以及 appActivity，因为要启用的是浏览器。首先，导入以下包。

```
import java.io.File;
import org.openqa.selenium.remote.DesiredCapabilities;
import io.appium.java_client.remote.AndroidMobileCapabilityType;
```

下面介绍如何测试移动 Web App 的 Desired Capabilities 配置，如代码清单 7-4 所示。

代码清单 7-4　　　　移动 Web App 的 Desired Capabilities 配置

```
DesiredCapabilities capabilities = new DesiredCapabilities();
//测试手机名为 Redmi 3
capabilities.setCapability("deviceName", "Redmi 3");
//选择 appium 作为自动化测试引擎
 capabilities.setCapability("automationName", AutomationName.APPIUM);
//待测移动设备为 Android 版本
 capabilities.setCapability("platformName", "Android");
//待测移动设备 Android 版本为 5.1.1
 capabilities.setCapability("platformVersion", "5.1.1");
//待测浏览器为 Chrome 浏览器
 capabilities.setCapability("browser","Chrome");
```

3. 连接 Appium 服务器

在 Desired Capabilities 配置完成后，需要初始化 Android Driver 来连接 Appium 服务器。以移动端 Web App 为例。首先，导入以下包。

```
import java.net.URL;
import io.appium.java_client.android.AndroidDriver;
```

初始化 Android Driver。

```
driver = new AndroidDriver(new URL("http://127.0.0.1:4723/wd/hub"),capabilities);
```

使用配置的 Desired Capabilities 启动待测设备中的 Chrome 浏览器，如代码清单 7-5 所示。

代码清单 7-5　　　　使用 Desired Capabilities 启动设备中的 Chrome 浏览器

```
package com.shijie.testScripts;

import java.net.MalformedURLException;
```

```java
import java.net.URL;
import java.util.concurrent.TimeUnit;

import io.appium.java_client.android.AndroidDriver;
import io.appium.java_client.remote.AutomationName;

import org.openqa.selenium.remote.DesiredCapabilities;
import org.testng.annotations.Test;
import org.testng.annotations.BeforeClass;
import org.testng.annotations.AfterClass;

public class TestApplication {
    private AndroidDriver driver;
 @Test
 public void testExample() {
     //测试脚本写在这里
 }
 @BeforeClass
 public void beforeClass() throws MalformedURLException {
     DesiredCapabilities capabilities = new DesiredCapabilities();
    capabilities.setCapability("deviceName", "Redmi 3");
    capabilities.setCapability("automationName", AutomationName.APPIUM);
    capabilities.setCapability("platformVersion", "5.1.1");
    capabilities.setCapability("platformName", "Android");
    capabilities.setCapability("browser","Chrome");
    driver = new AndroidDriver(new URL("http://127.0.0.1:4723/wd/hub"),capabilities);
    driver.manage().timeouts().implicitlyWait(30, TimeUnit.SECONDS);
 }


 @AfterClass
 public void afterClass() {
     driver.close();//关闭驱动,退出浏览器
 }
}
```

7.9 识别对象的 API 方法

Appium 是基于 GUI 的自动化测试，主要是围绕着界面的控件元素来进行的，所以编写测试脚本的第一步就是识别测试对象。使用 findElement 和 findElements 的原理基本相同，表示查找一个元素和查找一组元素。需要注意的是，使用 findElement 的返回结果只有一个，如果存在多个元素，则会定位到第一个。

Appium 提供了根据 name、className 和 id 等方法来查找控件，详细的内容见表 7-5。

表 7-5 对象识别方法

API 方法	方 法 描 述
findElementByName	通过控件的 Text 来查找控件，这个方法已经取消
findElementByClassName	通过空间的 ClassName 来查找控件
findElementById	通过控件的 ResourceId 来查找控件
findElementByAccessibilityId	通过空间的 Content Description 来查找控件
findElementByXPath	通过控件的 XPath 来查找控件
findElementByAndroidUIAutomator	通过 UI Automator 的定位方式来查找控件
findElementBycssSelector	通过 CSS Selector 来查找控件，只适用于 Web App 和混合 App
findElementByLinkText	通过链接的文本来查找控件，只适用于 Web App 和混合 App
findElementByPartialLinkText	通过链接的部分文本来查找控件，只适用于 Web App 和混合 App
findElementByTagName	通过 TagName 来查找控件，只适用于 Web App 和混合 App

7.9.1 通过 Name 属性识别

通过 Name 属性识别对象，在 Appium 1.6 以后的版本已经不支持。

示例如下所示。

```
el = driver.findElementByName("7");
assertThat(el.getText(),equalTo("7"));
```

尝试用 UI Automator Viewer，获取 name 属性值。属性列表中没有 name 属性，建议先尝试 text 属性，一般情况下都会成功，如图 7-46 所示。

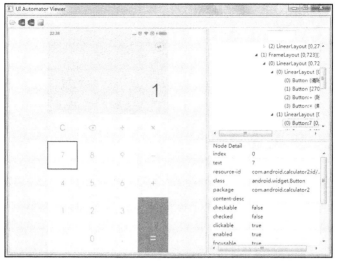

图 7-46　控件元素识别

采用下面代码可以进行替换。

```
WebElement prgLanguge = driver.findElementByAndroidUIAutomator("text(\"7\")");
```

7.9.2　通过 ClassName 属性识别

要通过 ClassName 属性识别对象,可以使用 UI Automator Viewer 工具直接查看,如图 7-47 所示。

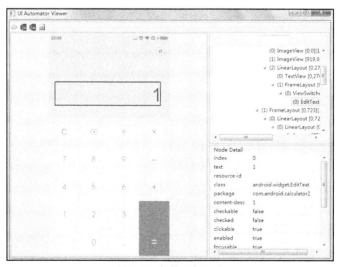

图 7-47　控件元素识别

如果对象的 ClassName 属性是唯一的，通过 ClassName 属性可以唯一识别对象。

```
el = driver.findElementByClassName("android.widget.EditText");
Assert.assertEquals(targetEle.getText(), "1");
```

如果对象的 ClassName 不唯一，又无法通过 name 和 id 来定位元素，则可以通过 ClassName 获得该类型的所有控件，然后根据元素的索引定位该控件，如下所示。

```
List<WebElement> lis = driver.findElementsByClassName("android.widget.Button");
WebElement targetEle = lis.get(0);
Assert.assertEquals(targetEle.getText(), "已定业务");
```

 使用这种方法因为系统要获取列表，执行效率较低。

7.9.3 通过 Id 属性识别

要通过 Id 属性识别对象，可以使用 UI Automator Viewer，如图 7-48 所示。

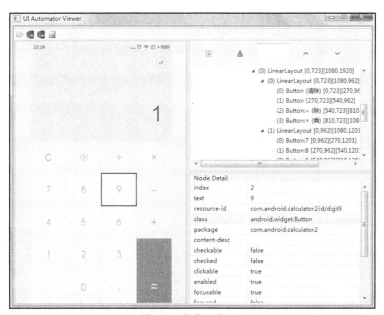

图 7-48 控件元素识别

resource-id 一般是唯一的，通过它可以唯一定位控件元素。

```
WebElement form = driver.findElementById("com.android.calculator2:id/digit9");
Assert.assertEquals(form.getText(), "9");
```

7.9.4 通过 AccessibilityId 识别

AccessibilityId 可以通过 UI Automator Viewer 或者 Appium Inspector 获得。在 Android 系统中，这个属性等同于 contentDescription。需要注意的是，通过 "Accessibility ID" 来定位元素，对于 Android 就是 content-description。这里直接定位，如图 7-49 所示。

图 7-49　菜单元素识别

因为本例中的 contentDescription，示例无法运行，所以假设该 "已定业务" 控件的 contentDescription 属性值为 text_order_query_description。

```
WebElement form = driver.findElementByAccessibilityId("菜单");
Assert.assertEquals(form.getText(), "菜单");
```

这个属性有助于一些生理功能有缺陷的人使用该应用程序。比如，有一个 ImageView 里面放置了一张颜色复杂的图片，可能一些色弱、色盲的人分不清这张图片中画的是什么东西。如果用户安装了辅助浏览工具 TalkBack，它就会大声朗读出用户目前正在浏览的内容。TextView 控件，TalkBack 可以直接读出里面的内容，但是对于 ImageView 控件，TalkBack 就只能去读 contentDescription 的值，告诉用户这个图片到底是什么。鉴于这是一个隐藏属性，而 Android 上用于查找控件的各种属性可能有所缺失或者重复（比如 id 重复，一个列表下面的所有项可能都叫作"id/text1"），所以最佳办法就是与开发团队沟通好，对于每个 View 都赋予一个唯一的 contentDescription。

7.9.5 通过 XPath 识别

XPath 是自动化测试定位中的重要方法，是非常强大的元素查找方式，使用这种方法几乎可以定位到页面上的任意元素。

1. 什么是 XPath

首先我们了解什么是 XPath。XPath 是 XML Path 的简称，因为 Android 定位文档本身就是一个特殊的 XML 页面，所以可以使用 XPath 的语法来定位页面元素。

现在以图 7-50 所示 XML 代码为例，假设要引用对应的对象，图中为相应的 XPath 语法。

图 7-50　控件元素识别

选中页面上 Chrome 浏览器的图标（即源码中高亮选中的一行），查看层级关系，1、2、

3、4、5 代表层级顺序，如图 7-51 所示。

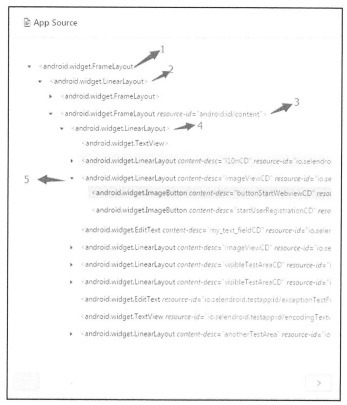

图 7-51　元素层级关系

2．XPath 绝对路径

针对图 7-51 所示的层级关系，获取第 5 层级下的第一个 ImageButton，绝对路径的写法只有以下一种。

```
/hierarchy/android.widget.FrameLayout/android.widget.LinearLayout/android.widget.
 FrameLayout[2]/android.widget.LinearLayout/android.widget.LinearLayout[2]/android.
 widget. ImageButton[1]
```

注意以下几点。

（1）元素的 XPath 绝对路径可通过 Appium Inspector 直接查询。

（2）一般不推荐使用绝对路径的写法，因为一旦页面结构发生变化，该路径也随之失效，必须重新写。

（3）路径偏长，而且因为只有 class 的值，对于一些页面控件较多的情况，可能不止一个绝对路径，所以这种写法也不一定唯一。

(4) 绝对路径以"/"表示，而下面要讲的相对路径则以"//"表示，这个区别非常重要。当 XPath 的路径以"/"开头时，表示让 XPath 解析引擎从文档的根节点开始解析。当 XPath 路径以"//"开头时，则表示让 XPath 引擎从文档中任意符合的元素节点开始进行解析。而当"/"出现在 XPath 路径中时，则表示寻找父节点的直接子节点；当"//"出现在 XPath 路径中时，表示寻找父节点下任意符合条件的子节点，不管嵌套了多少层级。弄清这个原则，用户就可以理解 XPath 路径可以将绝对路径和相对路径混合在一起来进行表示。

3．相对路径

要查找页面根元素，使用//。

要查找页面上所有的 input 元素，使用//input。

要查找页面上第一个 form 元素内的直接子 input 元素(即只包括 form 元素的下一级 input 元素，使用绝对路径表示，"/")，使用//form[1]/input。

要查找页面上第一个 form 元素内的所有子 input 元素(只要在 form 元素内的 input 都算，不管嵌套了多少个其他标签，使用相对路径表示，"//")，使用//form[1]//input。

要查找页面上的第一个 form 元素，使用//form[1]。

要查找页面上 id 为 loginForm 的 form 元素，使用//form[@id='loginForm']。

要查找页面上 name 属性为 username 的 input 元素，使用//input[@name='username']。

要查找页面上 id 为 loginForm 的 form 元素下的第一个 input 元素，使用//form[@id='loginForm']/input[1]。

要查找页面上 name 属性为 contiune 且 type 属性为 button 的 input 元素，使用//input[@name='continue'] [@type='button']。

要查找页面上 id 为 loginForm 的 form 元素下的第 4 个 input 元素，使用//form[@id='loginForm']/input[4]。

XPath 功能很强大，所以也可以写得更加复杂一些。查看上面的 App Source，它通过相对路径来进行 XPath 的选择。

通过 App Resource，我们知道 Chrome 控件的 content-desc 为"buttonStartWebviewCD"且元素类型为 ImageButton。

```
MobileElement chrome = driver.findElement(By.xpath("//android.widget.ImageButton
    [@content -desc='buttonStartWebviewCD']"));
```

这里值得注意的是以下两点。

（1）下标从 1 开始，而不是从 0 开始。

（2）和 Web 不一样的就是标签的取值，这里 class 的值等于 android.widget.ImageButton，而不是看到的标签 TextView 用 class 代替。

4．模糊匹配

前面讲的都是 XPath 中基于准确元素属性的定位，其实 XPath 作为定位神器也可以用于模糊匹配。比如"Show Progress Bar for a while"这个控件的 text 属性值为 Show Progress Bar for a while，其控件类型为 android.widget.Button。下面用模糊匹配来进行定位。

（1）用 contains 关键字进行定位的代码如下。

```
driver.findElement(By.xpath("//android.widget.Button [contains(@text, 'Show Progress')] "));
```

这句话的意思是寻找页面中 text 属性值包含 Show Progress 这个单词的所有 android.widget.Button 元素，@后面的 text 可以更换为元素任意的属性名。

（2）用 start-with 关键字进行定位的代码如下。

```
driver.findElement(By.xpath("//android.widget.Button[starts-with(@text,'Show Progress')]"));
```

这句话的意思是寻找 text 属性以"Show Progress"开头的 android.widget.Button 元素，其中@后面的 text 可以替换成元素的任意其他属性。

（3）用 Text 关键字进行定位的代码如下。

```
driver.findElement(By.xpath("//*[text()='Show Progress Bar for a while'])");
```

直接查找页面当中所有 text 值为"Show Progress Bar for a while"的元素，根本就不用知道它的控件类型。这种方法也经常用于纯文字的查找。

在使用 text 的时候，要避免使用输入框的默认值，因为当用户输入值之后，就没有这个 text 了，也就找不到路径了。

5．使用 XPath 轴进行元素定位

通过大众点评网，用户可以看到美食、电影/演出等，以上图标所有的属性值都是一样的，只有下面的文字不同，如图 7-52 所示。

首先，定位文字控件"//android.widget.TextView[@text='美食']"，找到父控件（/parent::android.widget.RelativeLayout）。然后，找到 class 为"android.widget.ImageView"的子控件（/android.widget.ImageView），连起来就是"//android.widget.TextView[@text='美食'] /parent::android.widget.RelativeLayout/ android.widget.ImageView"。父控件的位置可以用".."来代替，

很多人都知道"`..`"在路径里面指的就是上级。所以用户也可以用"//android.widget.TextView[@text='美食'] /../ android.widget.ImageView"来组成 XPath 路径。

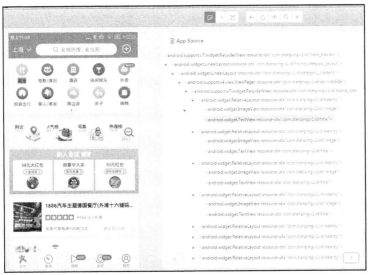

图 7-52　控件元素识别

XPath 中，"preceding-sibling::"可以找到前面的节点（也就是哥哥节点），"following-sibling::"可以找到后面的节点（也就是弟弟节点），所以可以使用文字控件(//android.widget.TextView[@text='美食'])直接找到哥哥节点，即图标控件(/preceding-sibling:: android.widget.ImageView)。合起来即为"//android.widget.TextView[@text='美食']//preceding- sibling:: android.widget.Image View"。

最后，关于 XPath 这种定位方式，AndroidDriver 会将整个页面的所有元素进行扫描以定位需要的元素，所以这是一个非常费时的操作。如果脚本中大量使用 XPath 做元素定位，将导致脚本执行速度大大降低。

7.9.6　通过 UIAutomator 识别

当通过 UIAutomator 识别对象时，关于 UIAutomator 如何识别对象可以参见具体的文档，这里不做介绍。Appium 底层封装了 UIAutomator，所以 Appium 也可以直接使用 UIAutomator 的识别方式。下面给出几个示例。

比如，要查找所有页面元素 class 为 android.widget.EditText 的控件，假设该控件的数量是 2，可以使用以下代码获取列表中第一个控件。

```
List<WebElement> editlist = (List<WebElement>) driver.findElementByAndroidUIAutomator("new
    Uiselector().className("+"android.widget.EditText"+")");
WebElement username = editlist.get(0);
Assert.assertEquals(username.getText(), "123456");
```

假设两个控件的 index 值不同,一个为 1,另一个为 2,可以使用以下代码获取 index 为 1 的控件。

```
List<WebElement> editlist = (List<WebElement>) driver.findElementByAndroidUIAutomator("new
    Uiselector().className("+"android.widget.EditText"+").index(1)");
WebElement username = editlist.get(0);
Assert.assertEquals(username.getText(), "123456");
```

假设该控件的数量为 5,其中 index 为 1 的控件有两个,可以使用以下代码获取其中一个可以单击的对象。

```
List<WebElement> editlist = (List<WebElement>) driver.findElementByAndroidUIAutomator("new
    Uiselector().className("+"android.widget.EditText"+").index(1).clickable(true)");
WebElement username = editlist.get(0);
Assert.assertEquals(username.getText(), "123456");
```

7.9.7　通过 cssSelector 识别

cssSelector 元素的定位方式与 XPath 比较类似,优点是执行速度快且比较稳定,主要针对移动 Web App 或者 Hybrid App 中的 Web 部分,如图 7-53 所示。

```
<html>
  <head>
    <title>Say Hello Demo</title>
  </head>
  <body>Hello, can you please tell me your name?
    <form name="myform" action="http://localhost:4450/sayhello" method="get">
      <div align="center">
        <br>
        <br>
        <input type="text" id="name_input" name="name" size="25" value="Enter your name here!">
        <br>
        <p>Prefered Car:
          <br>
          <select name="car">
            <option value="volvo">Volvo</option>
            <option value="mercedes">Mercedes</option>
            <option value="audi">Audi</option>
          </select>
        </p>
        <br>
        <input type="submit" value="Send me your name!">
        <br>
      </div>
    </form>
  </body>
</html>
```

图 7-53　selendroid-test-app-0.17.0.apk 中内嵌浏览器的 HTML 代码

cssSelector 常用的定位方式如下。

1. 根据 tagName 定位

通过以下方式定位 tagName 为 select 的元素。

CSS 定位表达式为 `select`。

Appium 定位表达式为 `driver.findElement(By.cssSelector("select"))`。

2. 根据 ID 定位

通过以下方式定位 id 为 "name_input" 的元素。

CSS 定位表达式为 `#name_input`。

Appium 定位表达式为：

```
driver.findElement(By.cssSelector("input#name_input"))  //html 标签和#id
driver.findElement(By.cssSelector("#name_input "))      //只是#id
```

3. 根据 className 定位

通过以下方式定位 classname 为 username 的元素。

CSS 定位表达式为 ".username"。

对于单一 class Appium 定位表达式为：

```
driver.findElement(By.cssSelector (".username"))   //只是 ".class"
```

对于复合 class Appium 定位表达式为：

```
driver.findElement(By.cssSelector (".username.**.***"))//复合class使用 ".classA" ".classB"
//进行叠加
```

4. 根据元素属性定位

（1）精准匹配。根据元素属性精准匹配的表达式如下。

```
driver.findElement(By.cssSelector("input[name= name_input]"))//属性名=属性值, id 和 class 等
//多种属性匹配顺序叠加
    driver.findElement(By.cssSelector("option [value= name_input '][type='text']"))//也是多属
//性叠加
```

（2）模糊匹配（正则表达式匹配属性）。比如在被测网页中，查找 www.b**d*.com 的链接。

CSS 定位表达式如下。

表达式1 a[href^='www.b**']

表达式2 a[href$='d*.com']

表达式3 a[href*='b**']

Appium 定位表达式分别如下。

driver.findElement(By.cssSelector(a[href^='www.b**']))，匹配到 a 头部。

driver.findElement(By.cssSelector(a[href$='d*.com']))，匹配到 a 尾部。

driver.findElement(By.cssSelector(at[href = 'b**']))，匹配到 a 中间。

（3）查找子元素。

①查找子元素。

要查找 select 下的 option 元素，可使用以下 CSS 定位表达式。

```
WebElement input= driver.findElement(By.cssSelector("f select>option[value='volvo']"));
```

②查找后代元素。

要查找 select 下的 option 元素，可使用以下 CSS 定位表达式。

```
WebElement input= driver.findElement(By.cssSelector("select option"));//搜索输入框
```

③查找第一个后代元素":first-child"。

```
WebElement span= driver.findElemet(By.cssSelector("select :first-child"));
```

冒号前要有空格，否则定位不到子页面元素。

④查找最后一个子元素":last-child"（类似于":first-child"）。

```
WebElement userName = driver.findEleme(By.cssSelector("select :last-child"));
```

⑤查找第 2 个子元素":nth-child(N)"（类似于":first-child"）。

```
WebElement userName = driver.findElemet(By.cssSelector("form#form :nth-child(2)"));
//定位到 form 下所有级别的第 2 个子元素
```

⑥要查找当前获取焦点的 input 页面元素，可以使用 focus。

```
WebElement userName = driver.findElemet(By.cssSelector("form#form :focus"));
```

⑦要查找可操作性的 input 页面元素，可以使用 enabled。

```
WebElement userName = driver.findElemet(By.cssSelector("form#form : enabled"));
```

⑧要查找处于勾选状态的 checkbox 页面元素，可以使用 checked。

```
WebElement userName = driver.findElemet(By.cssSelector("form#form : checked"));
```

⑨查询兄弟元素。

```
driver.findElement(By.cssSelector("input#name_input+span"));
```

上述代码表示定位到 id 属性值为 input_name 的同级 span 元素。

```
driver.findElement(By.cssSelector("input#name_input+br+span "));
```

上述代码表示定位到 id 属性值为 input_name 和 br 元素后面的同级 span 元素。

```
driver.findElement(By.cssSelector("input#name_input+*+span "));
```

上述代码表示定位到 id 属性值为 input_name 和某种元素类型后面的同级 span 元素，* 表示任意类型的页面元素。

7.9.8 通过 LinkText 识别

通过 LinkText 识别对象，定位网页中的超链接，需要 a 标签中的全部内容。

这个方法应该是针对 Webview 容器下面的控件定位的。

比如，针对 HTML 的代码如下。

```
<a href=http://news.baidu.com target=_blank class=mnav>新闻</a>
```

采用 findElementBylLinkText 的代码如下。

```
WebElement news = driver.findElementByLinkText("新闻");
```

7.9.9 通过 PartialLinkText 识别

该方法不适用于原生 App。通过 PartialLinkText 识别对象，定位网页中的超链接，需要 a 标签中的部分内容即可。

这个方法应该是针对 Webview 容器下面的控件来定位的。

比如，针对 HTML 的代码如下。

```
<a href=http://news.baidu.com target=_blank class=mnav>新闻</a>
```

采用 findElementByPartialLinkText 的代码如下。

```
WebElement news = driver.findElementByPartialLinkText("新");
```

7.9.10 通过 TagName 识别

该方法不适用于生态 App，通过 TagName 识别对象，主要用于匹配多个页面元素的情况，对查找的页面元素对象进行遍历、属性修改等。

比如，针对 HTML 的代码如下。

```
<a href=http://news.baidu.com target=_blank class=mnav>新闻</a>
```

采用 findElementByTagName 的代码如下。

```
WebElement news = driver. findElementByTagName ("a");
```

如果被测页面有多个链接元素，只有第一个被匹配的链接对象会返回并赋值给 news。如果要查找所有的链接元素，建议使用 findElementsByTagName 方法。

7.9.11 通过 by 类识别

在使用 Selenium Webdriver 进行元素定位时，在 PO 模式中，考虑到封装的方便性，通常使用 findElement 或 findElements 方法结合 By 类返回的元素句柄来定位元素。其中 By 类的常用定位方式共有以下 8 种。

- ById。
- ByLinkText。
- ByPartialLinkText。
- ByName。
- ByTagName。
- ByXPath。
- ByClassName。

- ByCssSelector。

比如，针对 HTML 的代码如下。

```
<a href=http://news.baidu.com target=_blank class=mnav>新闻</a>
```

使用 findElement 的代码如下。

```
WebElement news = driver. findElement(By.linkText("a"));
```

针对 HTML 的代码如下。

```
<a href=http://news.baidu.com target=_blank class=mnav>新闻</a>
```

使用 findElement 的代码如下。

```
WebElement news = driver. findElement(By.linkText("新闻"));
```

通过示例可以看出，findElement 仅是通过传入参数（识别方法）的途径进行元素定位，增加了识别的灵活性。

7.9.12 通过 getPageSource 识别

在混合 App 中，如果碰到 WebView 控件无法识别，那么 Chrome 浏览器插件定位的方式是定位到 WebView，但是在国内暂无法使用，因此可以使用 AppiumDriver 的 getPageSource 函数，把当前页面的元素以 XML 的方式显示出来。在 selendroid-test-app-0.17.0.apk 中内嵌浏览器时的 HTML 代码，如图 7-54 所示。

7.9.13 通过坐标界定对象识别

如果手机上的控件无法通过各种识别方式识别，只能采取坐标点的方式，用户可以获取控件的相对坐标，然后单击。

如代码清单 7-6 所示，为防止手机分辨率不同带来的影响，要避免使用固定的坐标，用以下方式可获取元素的坐标。

```html
<html>
  <head>
    <title>Say Hello Demo</title>
  </head>
  <body>Hello, can you please tell me your name?
    <form name="myform" action="http://localhost:4450/sayhello" method="get">
      <div align="center">
        <br>
        <br>
        <input type="text" id="name_input" name="name" size="25" value="Enter your name here!">
        <br>
        <p>Prefered Car:
          <br>
          <select name="car">
            <option value="volvo">Volvo</option>
            <option value="mercedes">Mercedes</option>
            <option value="audi">Audi</option>
          </select>
        </p>
        <br>
        <input type="submit" value="Send me your name!">
        <br>
      </div>
    </form>
  </body>
</html>
```

图 7-54　在 selendroid-test-app-0.17.0.apk 中内嵌浏览器时的 HTML 代码

代码清单 7-6　　　　　　　　　获取元素的坐标

```
//1.获取手机屏幕宽度
double Screen_X = driver.Manage().Window.Size.Width;
//2.获取手机屏幕高度
double Screen_Y = driver.Manage().Window.Size.Height;
//3.获取元素的起点坐标，即元素左上角点的横坐标
double startX = element.Location.X;
//4.获取元素的起点坐标，即元素左上角点的纵坐标
double startY = element.Location.Y;
//5.获取元素的宽度
double elementWidth = element.Size.Width;
//6.获取元素的高度
double elementHight = element.Size.Height;
```

根据以上取得的数据进行各种计算，因为屏幕的左上角点的坐标和屏幕的整体宽、高都是临时根据具体手机获取的，这样计算出来的元素坐标就不是固定的值了，所以上述代码对各种分辨率的屏幕的适配性就很灵活。

如代码清单 7-7 所示，假设页面上的所有控件都无法识别，要获得用户名和密码两个编辑框的相对坐标，并输入用户名 012，密码 345。

代码清单 7-7　　　　　　　　　　获得两个编辑框的相对坐标

```java
List<Point> coordinate = new ArrayList<Point>();
    coordinate = getWebelementPoint();
    driver.tap(0, coordinate.get(0).x, coordinate.get(0).y,10);
    driver.sendKeyEvent(7);
    driver.sendKeyEvent(8);
    driver.sendKeyEvent(9);
    driver.tap(0, coordinate.get(1).x, coordinate.get(1).y,10);
    driver.sendKeyEvent(10);
    driver.sendKeyEvent(11);
    driver.sendKeyEvent(12);
/*控件定义区
* 方法名：getWebelementPoint()
* 描述：通过类名识别编辑框对象，并将编辑框的坐标信息保存在数组中
* 返回值：坐标数组
* */
private List<Point> getWebelementPoint() {
    List<WebElement> editlist = (List<WebElement>) driver.findElementByAndroidUIAutomator("new Uiselector().className("+"android.widget.EditText"+")");
    List<Point> coord = new ArrayList<Point>();
    Point username = editlist.get(0).getLocation();
    Point password = editlist.get(1).getLocation();
    coord.add(username);
    coord.add(password);
    return coord;
}
```

7.9.14　按照权重识别

测试 App 分为两种，一种是本公司自主研发的 App，另一种是其他公司的 App。

针对自主研发的 App，建议使用 content-desc 字段来定位元素。这个字段在研发中对软件编写没有实际的用途，也不需要发生变化，要求开发人员加上该字段的值，方便我们在软

件开发完成前完成测试脚本的编写。

针对其他公司的 App，按照以下顺序，权重逐层递减。

（1）当页面元素有 id 属性时，最好尽量用 id 来定位。如果 id 属性存在重复，就只有选择其他定位方法。

（2）Xpath 功能很强大，但因为要全页面遍历，性能较差，尽量少用。如果确实少数元素不好定位，或者考虑跨平台需求可以选择 XPath，针对 Web 类型的 App 可以选择 cssSelector。

（3）针对 Web 类型的 App，当要定位一组元素中相同元素时，可以考虑用 tagName 或 name。

（4）针对 Web 类型的 App，当有链接需要定位时，可以考虑使用 linkText 或 partialLinkText 方式。

7.10　其他 API 方法详解

7.9 节介绍了控件对象识别的 API 方法。除了这些 API 方法外，Appium 提供的接口还包括手势操作、系统操作等。本节将对常见的 API 方法进行阐述。

7.10.1　与控件信息相关的 API 方法

识别控件对象后，需要对控件的信息进行验证。Appium 提供了一些专门的 API 方法用于获取控件信息，见表 7-6。

表 7-6　与控件信息相关的 API 方法

API 方法	方法描述
public String getText()	获取控件显示的文本信息。返回值为 String 类型，返回文本信息
public String getAttribute(String name)	获取控件的属性值，name 为属性字段，如 text、clickable、resource-id 等
public Point getLocation()	获取控件的位置信息，返回值为 Point 类型。返回控件的坐标值
public String getTagName()	获取控件的 tag 名称，不适用原生 App，返回值为 String 类型，返回 tag 名称

续表

API 方法	方法描述
public void click()	单击控件
public void clear()	如果是文本输入框,清空控件的文本
public boolean isEnabled()	判断控件是否处于可用状态,返回值为 Boolean 类型。如果显示则返回 true;如果不显示返回 false
public boolean isSelected()	判断控件是否被选中,返回值为 Boolean 类型。如果显示则返回 True;如果不显示返回 False
public boolean isDisplayed()	判断控件是否显示,返回值为 Boolean 类型。如果显示则返回 True;如果不显示返回 False
public void sendKeys(CharSequence…keysToSend)	模拟输入文本到控件中。keysToSend 表示输入的文本字符串
public void setValue(String value)	对控件的 value 属性重新赋值。value 表示待输入的文本字符

7.10.2 与手势相关的 API 方法

AndroidDriver、AppiumDriver 继承自 WebDriver,针对手机功能的特殊性,增加了一些专门用于操作手机的手势,包括滑动、缩放等。与手势相关的 API 方法见表 7-7。

表 7-7 与手势相关的 API 方法

API 方法	方法描述
public void swipe(int startx, int starty, int endx, int endy, int duration)	从屏幕的 A 点滑到 B 点 startx:起点 x 坐标 starty:起点 y 坐标 endx:终点 x 坐标 endy:终点 y 坐标 duration:持续时间
rotate	设置屏幕方向

7.10.3 与 TouchAction 相关的 API 方法

TouchAction 的原理是将一系列的动作放在一个链条中,然后将该链条传递给服务器。服务器接收到该链条后,解析各个动作,然后逐个执行,见表 7-8。

表 7-8　与系统环境或者系统硬件信息有关的 API 方法

API 方 法	方 法 描 述
public TouchAction press(WebElement el)	单击控件的中心位置，返回值为 TouchAction 对象。el 为要单击的控件
public TouchAction press(int x, int y)	单击坐标点的位置，返回值为 TouchAction 对象，x、y 为待单击的坐标值
public TouchAction press(WebElement el, int x, int y)	单击控件，返回值为 TouchAction 对象。el 为要单击的控件；x、y 为相对于控件左上角的偏移量
public TouchAction release()	释放操作
public TouchAction moveTo(WebElement el)	移动当前的触摸到指定控件的中心位置。el 为指定控件
public TouchAction moveTo(int x, int y)	移动到指定的坐标点。x、y 为相对于当前坐标的偏移量
public TouchAction moveTo(WebElement el, int x, int y)	移动当前的触摸到指定控件的位置。x、y 为相对于控件左上角的偏移量
public TouchAction tap(WebElement el)	单击控件的中心位置。el 为要单击的控件
public TouchAction tap(int x, int y)	单击坐标点的位置，返回值为 TouchAction 对象。x、y 为待单击的坐标值
public TouchAction tap(WebElement el, int x, int y)	单击控件的位置。el 为要单击的控件；x、y 为相对于控件左上角的偏移量
public TouchAction waitAction()	等待
public TouchAction waitAction(Duration duration)	等待指定的时间。duration 为等待时间
public TouchAction longPress(WebElement el)	按住控件不放。el 为待按控件
public TouchAction longPress(WebElement el, Duration duration)	按住控件不放，持续指定的时间。el 为待按控件；duration 为持续时间
public TouchAction longPress(int x, int y)	按住坐标点位置不放。x、y 为待长按的坐标值
public TouchAction longPress(int x, int y, Duration duration)	按住坐标点位置不放，持续指定的时间。x、y 为待长按的坐标值
public TouchAction longPress(WebElement el, int x, int y)	长按控件上指定位置。el 为要单击的控件。x、y 为相对于控件左上角的偏移量
public TouchAction longPress(WebElement el, int x, int y, Duration duration)	长按控件上指定位置，持续指定的时间。el 为要单击的控件，x、y 为相对于控件左上角的偏移量；duration 为持续时间
public void cancel()	取消操作
public TouchAction perform()	执行操作

7.10.4　与系统操作相关的 API 方法

与系统操作相关的 API 方法，见表 7-9。

表 7-9　与系统操作相关的 API 方法

API 方法	方法描述
public Set<String> getContextHandles()	获取当前会话所有可用的上下文(Context)，返回值为 Set<String> 类型。返回上下文的值
public String getContext()	获取当前会话正在使用的上下文，返回值为 String 类型。返回上下文的值
public WebDriver context(String name)	跳转到指定的上下文，name 为待跳转的上下文值。返回值为 WebDriver 类型
default void pressKeyCode(int key)	模拟发送一个硬件码给手机，key 为硬件码，具体参考 6.2.2 节
default void longPressKeyCode(int key)	模拟发送一个长按硬件码给手机，key 为硬件码，具体参考 6.2.2 节
default String currentActivity()	获取当前正在显示的 Activity 信息。返回值为 String 类型，返回值为 Activity 值
default String getCurrentPackage()	获取当前正在显示的 Package 信息。返回值为 String 类型，返回值为 Package 值
default byte[] pullFile(String remotePath)	拉取手机上的文件，返回值为以 Base64 格式编码的数据。在 Android 上，remotePath 为文件放置在手机上的位置
default byte[] pullFolder(String remotePath)	拉取手机上的文件夹，返回值为以 Base64 格式编码的数据。remotePath 为文件放置在手机上的位置
default void pushFile(String remotePath, byte[] base64Data)	将多个以 Base64 格式编码的数据推送到手机的文件路径
default void pushFile(String remotePath, File file)	将一个以 Base64 格式编码的数据推送到手机的文件路径
default void hideKeyboard()	如果有键盘出现，隐藏键盘
default void startActivity(Activity activity)	启动某个 Activity，如果 Activity 属于另外一个应用程序，则应用程序被打开
public void openNotifications()	打开消息通知栏
default Connection getConnection()	返回当前网络连接的类型，范围值为当前网络的连接类型
default void setConnection(Connection connection)	设置网络，connection 值为 NONE(0)、AIRPLANE(1)、WIFI(2)、DATA(4)、ALL(6)
Public byte[] getScreenshotAs(OutputType)	将手机屏幕截屏另存为计算机上的文件，保存文件类型
default void launchApp()	在测试设备上启动 Desired Capabilities 中指定的应用程序
default void installApp(String appPath)	安装 App
default boolean isAppInstalled(String bundleId)	判断 App 是否安装
default void resetApp()	重置当前运行的 App
default void runAppInBackground(Duration duration)	将被测程序放置在后台持续运行一段时间

7.11 Android 测试实例

目前主流应用程序大体分为 3 类,即移动 Web App、混合 App 和原生 App,如图 7-55 所示。本节结合实例介绍 Appium 中这 3 种应用程序的测试。

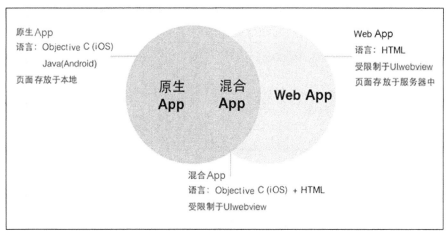

图 7-55 Android 应用程序

7.11.1 Android 原生 App 实例

原生 App 是基于客户端模式(Client-Side 模式)的应用程序(基于客户端的应用程序需要用 Android SDK 来开发,并且在用户的设备上安装一个以".apk"为扩展名的文件)。

以手机自带的计算器为例,执行的操作如下。

(1)打开计算器。

(2)输入 5。

(3)输入 3。

(4)验证结果是否为 8。

具体的操作步骤如下。

(1)将手机连接到计算机。

将手机设置为调试模式,通过 USB 连接计算机。为了测试 adb 是否已经成功连接,在命令行界面,输入以下命令。

```
adb devices
```

弹出结果,证明连接成功,如图 7-56 所示。

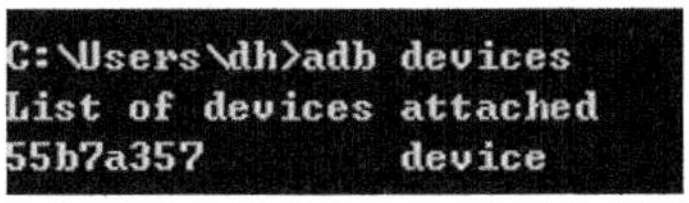

图 7-56 连接手机

(2)获取计算器 App 对应的包名和 Main Activity。输入以下命令。

```
adb shell dumpsys window w |findstr \/ |findstr name=
```

执行结果,如图 7-57 所示。

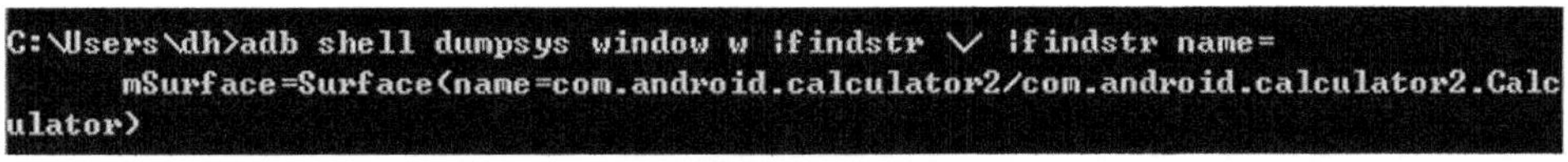

图 7-57 获取包名和 Main Activity

- 包名:com.android.calculator2。
- Main Actvity:com.android.calculator2.Calculator。

(3)获取手机信息,如图 7-58 所示。

- deviceName:REDMI NOTE 3。
- platformVersion:5.1.1。

(4)设置 DesiredCapabilities。根据前面获取到的信息,配置 DesiredCapabilities,如代码清单 7-8 所示。

(5)获取控件元素定位方式。启动 UI Automator Viewer 查看页面对象。UI Automator Viewer 位于 Android SDK 安装目录下的 tools 目录下,如图 7-59 所示。

图 7-58 查看手机版本

代码清单 7-8　　　　　　　DesiredCapabilities 的配置

```
DesiredCapabilities capabilities = new DesiredCapabilities();
capabilities.setCapability("deviceName", "REDMI NOTE 3");
capabilities.setCapability("platformVersion", "5.1.1");
```

```
capabilities.setCapability("platformName", "Android");
capabilities.setCapability("appPackage", "com.android.calculator2");
capabilities.setCapability("appActivity", ".Calculator");
capabilities.setCapability("noSign", "True");
```

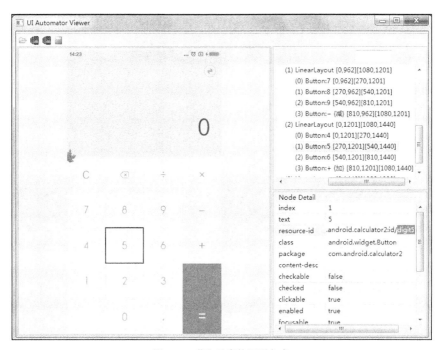

图 7-59　获取元素的识别方式

其中，5 的 id 为 digit5；+的 id 为 plus；3 的 id 为 digit3；=的 id 为 equal；输出界面的 class 为 android.widget.EditText。

（6）查找程序页面元素，输入程序所需要的字段，如代码清单 7-9 所示。

代码清单 7-9　　　　　　　　查找程序页面元素

```
WebElement digit3 = driver.findElementById("com.android.calculator2:id/digit3");
WebElement plus = driver.findElementById("com.android.calculator2:id/plus");
WebElement digit5 = driver.findElementById("com.android.calculator2:id/digit5");
WebElement equal= driver.findElementById("com.android.calculator2:id/equal");
WebElement result= driver.findElementByClassName("android.widget.EditText");
```

（7）操作对象并做断言，如代码清单 7-10 所示。

代码清单 7-10　　　　　　　　　　　操作对象并做断言

```
digit3.click();
        plus.click();
        digit5.click();
        equal.click();
        int actualResult = Integer.parseInt(result.getText());
        Assert.assertEquals(actualResult, 8);
```

（8）完整的测试代码如代码清单 7-11 所示。

代码清单 7-11　　　　　　　　　　　完整代码

```
package com.baidu.test.baidu;
import java.net.URL;
import io.appium.java_client.android.AndroidDriver;
import io.appium.java_client.android.AndroidElement;
import org.openqa.selenium.WebElement;
import org.openqa.selenium.remote.DesiredCapabilities;
import org.testng.Assert;
import org.testng.annotations.AfterTest;
import org.testng.annotations.BeforeTest;
import org.testng.annotations.Test;

public class NativeAppTest {
    AndroidDriver<AndroidElement> driver;

    @BeforeTest
    public void start() throws Exception  {
        DesiredCapabilities capabilities = new DesiredCapabilities();
        capabilities.setCapability("deviceName", "REDMI NOTE 3");
        capabilities.setCapability("platformVersion", "5.1.1");
        capabilities.setCapability("platformName", "Android");
        capabilities.setCapability("appPackage", "com.android.calculator2");
        capabilities.setCapability("appActivity", ".Calculator");
        capabilities.setCapability("noSign", "True");
```

```java
        driver = new AndroidDriver<>(new URL("http://127.0.0.1:4723/wd/hub"), capabilities);
    }

    @Test
    public void testWebApp(){
        //识别测试对象
        WebElement digit3 = driver.findElementById("com.android.calculator2:id/digit3");
        WebElement plus = driver.findElementById("com.android.calculator2:id/plus");
        WebElement digit5 = driver.findElementById("com.android.calculator2:id/digit5");
        WebElement equal= driver.findElementById("com.android.calculator2:id/equal");
        WebElement result= driver.findElementByClassName("android.widget.EditText");
        //执行测试用例
        digit3.click();
        plus.click();
        digit5.click();
        equal.click();
        int actualResult = Integer.parseInt(result.getText());
        Assert.assertEquals(actualResult, 8);
    }

    @AfterTest
    public void stop()
    {
        driver.quit();
    }
}
```

7.11.2　Android 移动 Web App 实例

基于浏览器的 Web 应用程序（基于浏览器的应用程序开发需要遵循 Web 标准，通过一个 Web 浏览器来访问开发的应用程序，不需要在用户的设备上安装其他任何程序），类似于现在所说的轻应用，是触屏版的网页应用。浏览器应用实际上就是普通网站，但是针对 Android 等移动设备做的浏览器进行了特殊的优化。

笔者曾经指导过一个自动化测试项目，它不复杂，但是很典型，要求对所有过期页面加上弹窗，告知用户页面已经过期，并引导用户浏览新的页面。

要求如下。

（1）涉及 1210 个页面。

（2）考虑到手机的兼容性，涉及 30 多个机型。

（3）操作系统涉及 Android 和 iOS。

这个时候如果用手工测试，那么工作量就很大，因此最后决定采用 Appium。本例进行了简化，以其中一种机型为例。具体的操作步骤如下。

（1）将手机连接到计算机。

（2）获取手机信息。

- deviceName：REDMI NOTE 3。

- platformVersion：5.1.1。

（3）对象识别。通过 Chrome 浏览器模拟手机客户端进行对象识别，结果如图 7-60 所示。

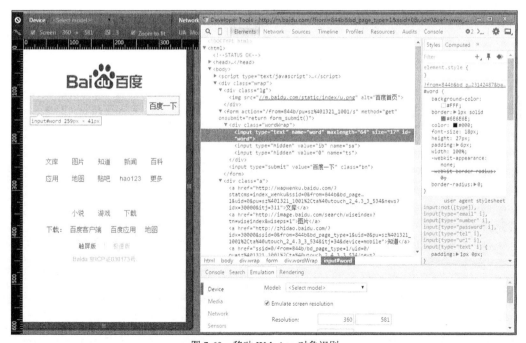

图 7-60　移动 Web App 对象识别

弹出框的 id 为 g-pc-modal-main。

(4) 设置 DesiredCapabilities, 用来启动 Chrome 浏览器, 如代码清单 7-12 所示。

代码清单 7-12　　　　　　　　DesiredCapabilities 的配置

```
DesiredCapabilities capabilities = new DesiredCapabilities();
capabilities.setCapability("deviceName", "REDDMI NOTE 3");
capabilities.setCapability("platformVersion", "5.1.1");//
capabilities.setCapability("platformName", "Android");//
capabilities.setCapability("browserName", "Chrome");
driver = new AndroidDriver<>(new URL("http://127.0.0.1:4723/wd/hub"), capabilities);
```

(5) 导航到指定页面, 如代码清单 7-13 所示。

代码清单 7-13　　　　　　　　导航到指定页面

```
String baseUrl = "https://www.####.com/";

    driver.manage().timeouts().implicitlyWait(30, TimeUnit.SECONDS);
    driver.get(baseUrl + "/");
```

(6) 通过 id 判断弹出框是否存在。

```
driver.findElementById("g-pc-modal-main").isDisplayed();
```

(7) 实现截图, 如代码清单 7-14 所示。

代码清单 7-14　　　　　　　　实现截图

```
    public static void takeScreenShot(AndroidDriver driver, String baseUrl)
    /*
    * 截图*/
      {
        File screenShotFile = ((TakesScreenshot)driver).getScreenshotAs(OutputType.FILE);
        try {
          FileUtils.copyFile(screenShotFile, new File("D:\\AutoScreenCapture\\" +
            setScreenShotName(baseUrl)+ ".jpg"));
        }
        catch (IOException e) {e.printStackTrace();}
      }
```

(8) 使用 TestNG 测试框架运行测试程序, 在 BeforeTest 中对启动参数进行配置, 在

AfterTest 中清理环境。完整代码如代码清单 7-15 所示。

代码清单 7-15　　　　　　　　完整代码

```java
package com.baidu.test.baidu;
import java.io.File;
import java.io.IOException;
import java.net.URL;
import java.util.concurrent.TimeUnit;
import io.appium.java_client.android.AndroidDriver;
import io.appium.java_client.android.AndroidElement;
import org.apache.commons.io.FileUtils;
import org.openqa.selenium.OutputType;
import org.openqa.selenium.TakesScreenshot;
import org.openqa.selenium.remote.DesiredCapabilities;
import org.testng.annotations.AfterTest;
import org.testng.annotations.BeforeTest;
import org.testng.annotations.Test;




public class WebAppTest {
    AndroidDriver<AndroidElement> driver;


    @BeforeTest
    public void start() throws Exception {
        // TODO Auto-generated method stub
        DesiredCapabilities capabilities = new DesiredCapabilities();
        capabilities.setCapability("deviceName", "REDDMI NOTE 3");
        capabilities.setCapability("platformVersion", "5.1.1");//
        capabilities.setCapability("platformName", "Android");//
        capabilities.setCapability("browserName", "Chrome");
        driver = new AndroidDriver<>(new URL("http://127.0.0.1:4723/wd/hub"),
            capabilities);
    }
```

```java
@Test
public void testWebApp(){
    String baseUrl = "https://www.####.com/";
    
    driver.manage().timeouts().implicitlyWait(30, TimeUnit.SECONDS);
    driver.get(baseUrl + "/");
    try {
        driver.findElementById("g-pc-modal-main").isDisplayed();
        System.out.printf("页面%S 找到弹出框:", baseUrl);
        }
    catch(Exception e){
        System.out.printf("页面%S 没有找到弹出框:", baseUrl);
        }
    finally{
        takeScreenShot(driver, baseUrl);
        }
}

public static void takeScreenShot(AndroidDriver driver, String baseUrl)
/*
 * 截图*/
    {
        File screenShotFile = ((TakesScreenshot)driver).getScreenshotAs(OutputType.FILE);
        try {
            FileUtils.copyFile(screenShotFile, new File("D:\\AutoScreenCapture\\" +
              setScreenShotName(baseUrl)+ ".jpg"));
            }
        catch (IOException e) {e.printStackTrace();}
    }

public static String setScreenShotName(String baseUrl)
/*
 * 截图地址以链接最后一个字段命名*/
    {
        String[] file = baseUrl.split("\\.");
```

```
            String filename = file[file.length - 1];
            return filename;
    }


    @AfterTest
    public void stop()
    {
        driver.quit();
    }
}
```

需要注意的是，在测试移动 Web App 时，需要使用 chromedriver。Appium 中自带有该驱动。位置在 Appium 安装目录\resources\app\node_modules\appium\node_modules\appium-chromedriver\chromedriver\win，如图 7-61 所示。

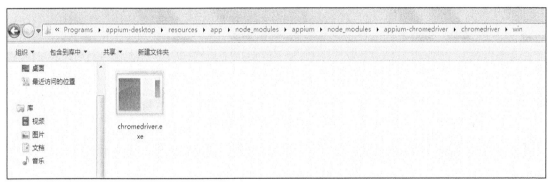

图 7-61　chromedriver 的位置

在测试过程中要求 chromedriver 与 Chrome 版本保持一定的映射关系。如果不一致，可以备份原有的版本，并将最新的版本放置在这个位置。

7.11.3　Android 混合 App 实例

用户通过两种方式访问 Web 内容，一种是通过传统的浏览器方式，另一种则是通过在 Android 应用程序的布局文件中包含一个 WebView 组件的方式来实现，如图 7-62 所示。

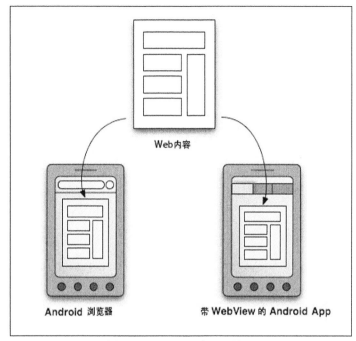

图 7-62　Android 混合 App 图示

有时想要在 Android 应用内部显示网页，但是不想用 Intent 对象打开浏览器进行浏览，如果要显示的网页保持在应用的资源里，或者想监听用户在网页上的操作，可以使用 WebView 控件，WebView 与 Android 内置的浏览器一样，都是使用 WebKit 引擎来显示网页的，并提供了在浏览记录中前进、后退、缩放页面和搜索文本等功能。

下面以 selendroid-test-app-0.17.0.apk 为例，APK 下载地址见 selendroid 官网，具体操作如下。

（1）单击 Google 浏览器图标，如图 7-63 所示。

（2）跳转到该页面，如图 7-64 所示。

（3）根据提示进行输入。通过以下步骤，实现上述功能。

①将手机连接到计算机。

②获取手机信息。

- deviceName：REDMI NOTE 3。
- platformVersion：5.1.1。

图 7-63 单击 Google 浏览器图标

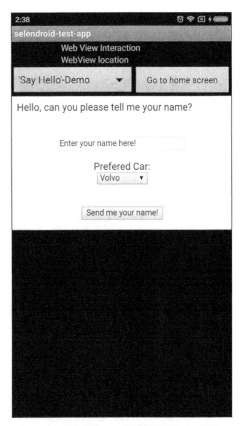
图 7-64 单击 Google 浏览器图标后页面跳转

③配置 DesiredCapabilities，配置项见表 7-10。

表 7-10　DesiredCapabilities 配置

配　置　项	配　置　值
Application Path	F:/ selendroid-test-app-0.17.0.apk
Platform Name	Android
Automation Name	Appium
Platform Version	5.1.1
Device Name	REDMI NOTE 3

④定位元素。

首先，定位 Google 浏览器图标字段，如图 7-65 所示。

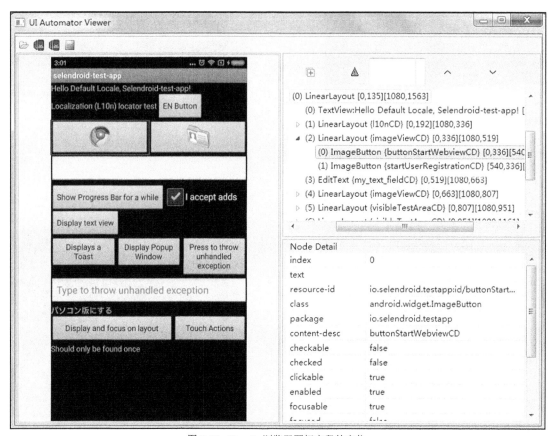

图 7-65　Google 浏览器图标字段的定位

Google 图标的 id 为 io.selendroid.testapp:id/buttonStartWebview。

然后，定位 Google 图标字段。

```
WebElement google = driver.findElementById("io.selendroid.testapp:id/buttonStartWebview ");
```

接下来，单击 Google 图标。

```
WebElement google = driver.findElementById("io.selendroid.testapp:id/buttonStartWebview ");
```

之后，定位页面跳转后的元素，如图 7-66 所示。

通过 UI Automator Viewer 可以看出，这是个 WebView 控件，UI Automator Viewer 无法进行定位。因此，通过代码查看页面是否包含 WebView 控件。

⑤查看界面是否包含 WebView，如代码清单 7-16 所示。

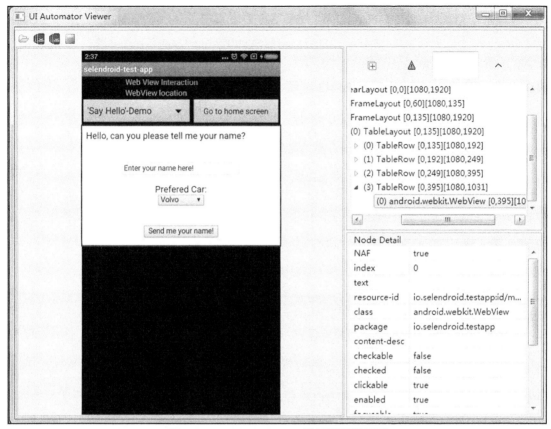

图 7-66　UI Automator Viewer 跳转到的页面元素

代码清单 7-16　　　　　　　　查看界面是否包含 WebView

```
try {contexts
    Set<String> contextNames = driver.getContextHandles();
    for (String contextName : contextNames) {
    // 用于判断被测 App 是 NATIVE_APP 还是 WEBVIEW，如果同时属于两者，就是混合 App
    System.out.println(contextName);
        }
```

上述代码返回一个 contexts 列表，如 NATIVE_APP 或 WEBVIEW_0，或者以 WEBVIEW 开头的字符，这返回的结果表示混合 App。

⑥界面包含 WebView，进行切换。

```
// 让 Appium 切换到 WebView 模式以便查找 Web 元素
driver.context("WEBVIEW_0");
```

使用所访问 context 的 id(Appium 处于这种会话模式，即所有命令解释为用于自动化 WebView，而不是原生 App。例如，当运行 getElementByTagName 时，它将操作 WebView 的 DOM，而不是返回 UIAElements。当然，某些 WebDriver 方法只在一个上下文（如 Native App 或者 WebView）中使用，如果环境错误，系统会提示报错)。

⑦输出 WebView 的页面元素。

```
System.out.println(driver.getPageSource());
```

⑧WebView 对象元素的对象识别。在代码中切换到 WebView 后使用 Driver，可以通过 GetPageSource()得到页面的元素并保存在一个文件里。通过上下文，把被测 App 识别为混合 App，切换模式后通过 getPageSource 获得整个页面的 XML 文件，通过查看该文件进行对象的识别，如代码清单 7-17 所示。

代码清单 7-17　　　　　　　　　对象识别

```
try {
    Set<String> contextNames = driver.getContextHandles();
    for (String contextName : contextNames) {
        System.out.println(contextName);
        // 用于判断被测 App 是 NATIVE_APP 还是 WEBVIEW，如果同时属于两者，就是混合 App
    }

    Thread.sleep(5000);// 设置延时
    driver.context("WEBVIEW_0");
    // 让 Appium 切换到 WebView 模式以便查找 Web 元素
    System.out.println(driver.getPageSource());
} catch (InterruptedException e) {
    e.printStackTrace();
}
```

⑨编写代码，如代码清单 7-18 所示。

代码清单 7-18　　　　　　　　　完整代码

```
package appiumsample;
```

```java
import io.appium.java_client.android.AndroidDriver;

import java.net.MalformedURLException;
import java.net.URL;
import java.util.Set;

import org.junit.After;
import org.junit.Assert;
import org.junit.Before;
import org.junit.Test;
import org.openqa.selenium.By;
import org.openqa.selenium.WebElement;
import org.openqa.selenium.remote.DesiredCapabilities;

public class testwebview {
    private AndroidDriver driver;
    @Before
    public void start() throws MalformedURLException {
        DesiredCapabilities capabilities = new DesiredCapabilities();
        driver = new AndroidDriver(new URL("http://127.0.0.1:4723/wd/hub"),capabilities);
    }


    @Test
    public void main() {

        WebElement chrome = driver.findElementid("buttonStartWebview"));
        chrome.click();
        //输出为 NATIVE_APP WEBVIEW_0，进行上下文模式切换
        try {
            Set<String> contextNames = driver.getContextHandles();
            for (String contextName : contextNames) {
                // 用于判断被测 App 是 NATIVE_APP 还是 WebView，如果同时属于两者，就是混合 App
                System.out.println(contextName);
            }
```

```java
        Thread.sleep(5000);
        // 让 Appium 切换到 WebView 模式以便查找 Web 元素
        driver.context("WEBVIEW_0");
        WebElement name = driver.findElementById("name_input");
        name.clear();
        WebElement car = driver.findElementByName("car").findElements(By.tagName("option")).
            get(1);
        name.sendKeys("bree");
         car.click();
        WebElement submit = driver.findElementsByTagName("input").get(1);
        submit.click();
        Thread.sleep(5000);// 设置延时
        Assert.assertEquals("Hello: bree", driver.getTitle());
    } catch (InterruptedException e) {
        e.printStackTrace();
    }
}


@After
public void stop() {
    driver.quit();
}
}
```

7.12　查看 Appium 日志

　　Appium 在运行的过程中，Appium 服务器一直输出各种日志。这些日志很清楚地显示一个测试用例从启动到完成的整个活动内容。熟悉这些日志的内容，有助于更好地理解整个测试流程，从而更好地定位程序运行中出现的各种问题。示例代码如代码清单 7-19 所示。

代码清单 7-19   日志示例代码

```java
package com.shijie.testScripts;

import io.appium.java_client.android.AndroidDriver;
import java.net.MalformedURLException;
import java.net.URL;
import org.openqa.selenium.WebElement;
import org.openqa.selenium.remote.DesiredCapabilities;
import org.testng.Assert;
import org.testng.annotations.AfterMethod;
import org.testng.annotations.BeforeMethod;
import org.testng.annotations.Test;

public class TestApplication {
    //定义变量
    private AndroidDriver driver;

    @BeforeMethod
    public void start() throws MalformedURLException {
        // App 地址
        String apppath = "F:\\selendroid-test-app-0.17.0.apk";
        // 配置 AndroidDriver
        DesiredCapabilities capabilities = new DesiredCapabilities();
        capabilities.setCapability("deviceName", "REDMI NOTE 3");
        capabilities.setCapability("platformVersion", "5.1.1");
        capabilities.setCapability("platformName", "Android");
        capabilities.setCapability("app", apppath);
        capabilities.setCapability("appPackage", "io.selendroid.testapp");
        capabilities.setCapability("appActivity", ".HomeScreenActivity");
        driver = new AndroidDriver(new URL("http://127.0.0.1:4723/wd/hub"),capabilities);
    }

    @Test
```

```java
public void main() throws InterruptedException {

    WebElement startRegister_btn = driver.findElementById("io.selendroid.testapp:id/
    startUserRegistration");
    //单击,页面跳转
    startRegister_btn.click();
    //用户名对象
    WebElement username_txt = driver.findElementById("io.selendroid.testapp:id/
    inputUsername");
    username_txt.sendKeys("shijie");
    //E-Mail对象
    WebElement email_txt = driver.findElementById("io.selendroid.testapp:id/
    inputEmail");
    email_txt.sendKeys("shijie@126.com");
    //密码对象
    WebElement password_txt = driver.findElementById("io.selendroid.testapp:id/
    inputPassword");
    password_txt.sendKeys("123456");
    WebElement name_txt=driver.findElementById("io.selendroid.testapp:id/inputName");
    name_txt.clear();
    name_txt.sendKeys("daming");
    //driver.hideKeyboard();
    WebElement languge_sel = driver.findElementById("io.selendroid.testapp:id/input_
    preferedProgrammingLanguage");
    languge_sel.click();
    WebElement prgLanguge = driver.findElementByAndroidUIAutomator("text(\"Scala\")");
    prgLanguge.click();
    WebElement accept_check = driver.findElementById("io.selendroid.testapp:id/
    input_adds");
    accept_check.click();
    WebElement register_btn = driver.findElementById("io.selendroid.testapp:id/
    btnRegisterUser");
    register_btn.click();
    WebElement label_name_data = driver.findElementById("label_name_data");
    Assert.assertEquals(label_name_data.getText().toString(), "daming");
}
```

```
    @AfterMethod
    public void stop() {
        driver.quit();
    }
}
```

在代码清单 7-20 所示的 Appium 服务器日志中，详细解释各命令的作用。

Appium 服务器日志等级分为以下几类。

- Message，简单的 Appium 服务器日志。

- [debug]message，Appium 服务器日志，等级为非 Debug。

- [debug] [AndroidBootstrap] [BOOTSTRAP LOG] [debug]message，Bootstrap 返回的日志。

- [debug] [UIAUTOMATOR] message，UiAutomator 执行返回的日志。

- [debug] [ADB] message，adb 命令执行返回的日志。

代码清单 7-20   日志详解

```
//启动 Appium 服务器成功

[info] [Appium] Welcome to Appium v1.7.0

[info] [HTTP] --> POST /wd/hub/session {"capabilities":[{"desiredCapabilities":{"app":"F:
\\selendroid-test-app-0.17.0.apk","appPackage":"io.selendroid.testapp","appActivity":".
HomeScreenActivity","platformVersion":"5.1.1","platformName":"Android","deviceName":"RE
DMI NOTE 3"}},{"requiredCapabilities":{}}],"desiredCapabilities":{"app":"F:\\selendroid-
test-app-0.17.0.apk","appPackage":"io.selendroid.testapp","appActivity":".HomeScreenActiv
ity","platformVersion":"5.1.1","platformName":"Android","deviceName":"REDMI NOTE 3"},"required
Capabilities":{}}

//收到从脚本传递的 post 请求，将 desiredCapabilities 参数全部传递给服务器端

[debug] [MJSONWP] Calling AppiumDriver.createSession() with args: [{"app":"F:\\selendroid-
test-app-0.17.0.apk","appPackage":"io.selendroid.testapp","appActivity":".HomeScreenAct
ivity","platformVersion":"5.1.1","platformName":"Android","deviceName":"REDMI      NOTE
3"},{},[{"desiredCapabilities":{"app":"F:\\selendroid-test-app-0.17.0.apk","appPackage":
"io.selendroid.testapp","appActivity":".HomeScreenActivity","platformVersion":"5.1.1","
platformName":"Android","deviceName":"REDMI NOTE 3"}},{"requiredCapabilities":{}}]]

[debug] [BaseDriver] Event 'newSessionRequested' logged at 1511922061565 (10:21:01 GMT+0800
(中国标准时间))
```

```
//创建session
[info] [Appium] Creating new AndroidDriver (v1.26.5) session
[info] [Appium] Capabilities:
[info] [Appium]   app: 'F:\\selendroid-test-app-0.17.0.apk'
[info] [Appium]   appPackage: 'io.selendroid.testapp'
[info] [Appium]   appActivity: '.HomeScreenActivity'
[info] [Appium]   platformVersion: '5.1.1'
[info] [Appium]   platformName: 'Android'
[info] [Appium]   deviceName: 'REDMI NOTE 3'
//生成的sessionid
[info] [BaseDriver] Session created with session id: bc8415e7-b66f-49ac-a966-e25a4e2684c0
//获取Java版本
[debug] [AndroidDriver] Getting Java version[info] [AndroidDriver] Java version is: 1.8.0_144
//检查adb是否存在，通过配置的环境变量ANDROID_HOME从固定路径下查找
[info] [ADB] Checking whether adb is present
[info] [ADB] Using adb.exe from D:\Android\android-sdk\platform-tools\adb.exe
//通过adb devices命令，查找当前连接的设备
[info] [AndroidDriver] Retrieving device list
[debug] [ADB] Trying to find a connected android device
[debug] [ADB] Getting connected devices...[debug] [ADB] 1 device(s) connected
//判断找到的设备中是否有与Desired Capabilities提供的platformName、platformVersion一致的设备。
//本例中查找Android操作平台，Android版本为5.1.1的待测设备
[info] [AndroidDriver] Looking for a device with Android '5.1.1'
[debug] [ADB] Setting device id to c1aeae297d72
[info] [ADB] Getting device platform version
[debug] [ADB] Getting connected devices...
[debug] [ADB] 1 device(s) connected
[debug] [ADB] Running 'D:\Android\android-sdk\platform-tools\adb.exe' with args: ["-P",5037,"-s","c1aeae297d72","shell","getprop","ro.build.version.release"][debug] [ADB] Current device property 'ro.build.version.release': 5.1.1
//找到设备并将该设备的udid设置为要使用的
[info] [AndroidDriver] Using device: c1aeae297d72
[info] [ADB] Checking whether adb is present
[info] [ADB] Using adb.exe from D:\Android\android-sdk\platform-tools\adb.exe
```

[debug] [ADB] Setting device id to c1aeae297d72

//要安装本地应用F:\selendroid-test-app-0.17.0.apk，先检查这个App是否实际存在，若存在则无须安装，
//若不存在则安装该应用

[info] [BaseDriver] Using local app 'F:\selendroid-test-app-0.17.0.apk'

[debug] [AndroidDriver] Checking whether app is actually present

[info] [AndroidDriver] Starting Android session

//通过adb shell echo ping，检查设备是否能够正常响应adb命令。若设备能够正常响应并输出"ping"则认为
//设备正常，这里响应有个超时时间（默认是5s）

[debug] [ADB] Running 'D:\Android\android-sdk\platform-tools\adb.exe' with args:
["-P",5037,"-s","c1aeae297d72","wait-for-device"]

[debug] [ADB] Getting connected devices...

[debug] [ADB] 1 device(s) connected

//通过adb shell echo ping 检查设备状态

[debug] [ADB] Running 'D:\Android\android-sdk\platform-tools\adb.exe' with args:
["-P",5037,"-s","c1aeae297d72","shell","echo","ping"]

//安装Appium Settings App，辅助设置设备的网络。安装前先判断是否已经存在于设备上

[debug] [Logcat] Starting logcat capture[debug] [AndroidDriver] Pushing settings apk to device...

[debug] [ADB] Getting install status for io.appium.settings

[debug] [ADB] Getting connected devices...

[debug] [ADB] 1 device(s) connected

//检查Appium Setting App是否安装成功

[debug] [ADB] Running 'D:\Android\android-sdk\platform-tools\adb.exe' with args:
["-P",5037,"-s","c1aeae297d72","shell","pm","list","packages","io.appium.settings"]

[debug] [ADB] App is installed

[debug] [ADB] Getting package info for io.appium.settings

[debug] [ADB] Getting connected devices...

[debug] [ADB] 1 device(s) connected

[debug] [ADB] Running 'D:\Android\android-sdk\platform-tools\adb.exe' with args:
["-P",5037,"-s","c1aeae297d72","shell","dumpsys","package","io.appium.settings"]

//检查appt是否到位

[info] [ADB] Checking whether aapt is present

[info] [ADB] Using aapt.exe from D:\Android\android-sdk\build-tools\19.1.0\aapt.exe[debug]
[ADB] The installed "io.appium.settings" package does not require upgrade (4 >= 4)

[debug] [ADB] Getting connected devices...[debug] [ADB] 1 device(s) connected

//获得待测设备的 Android API level

[debug] [ADB] Running 'D:\Android\android-sdk\platform-tools\adb.exe' with args: ["-P",5037,"-s","c1aeae297d72","shell","getprop","ro.build.version.sdk"]

[debug] [ADB] Current device property 'ro.build.version.sdk': 22

[debug] [ADB] Device API level: 22

[debug] [ADB] Getting connected devices...

[debug] [ADB] 1 device(s) connected

[debug] [ADB] Running 'D:\Android\android-sdk\platform-tools\adb.exe' with args: ["-P",5037,"-s","c1aeae297d72","shell","dumpsys","package","io.appium.settings"]

[debug] [ADB] Getting connected devices...[debug] [ADB] 1 device(s) connected

[debug] [ADB] Running 'D:\Android\android-sdk\platform-tools\adb.exe' with args: ["-P",5037,"-s","c1aeae297d72","shell","ps"]

[debug] [ADB] Device API level: 22

[debug] [ADB] Getting connected devices...

[debug] [ADB] 1 device(s) connected

//启动 Appium Setting App

[debug] [ADB] Running 'D:\Android\android-sdk\platform-tools\adb.exe' with args: ["-P",5037,"-s","c1aeae297d72","shell","am","start","-W","-n","io.appium.settings/.Settings","-a","android.intent.action.MAIN","-c","android.intent.category.LAUNCHER","-f","0x10200000"][debug] [ADB] Device API level: 22

[debug] [ADB] Getting connected devices...

[debug] [ADB] 1 device(s) connected

[debug] [ADB] Running 'D:\Android\android-sdk\platform-tools\adb.exe' with args: ["-P",5037,"-s","c1aeae297d72","shell","settings","put","secure","mock_location","1"]

//安装 Unlock App, 辅助设备解锁

[debug] [AndroidDriver] Pushing unlock helper app to device...

[debug] [ADB] Running 'D:\Android\android-sdk\platform-tools\adb.exe' with args: ["-P",5037,"-s","c1aeae297d72","install","C:\\Users\\dh\\AppData\\Local\\Programs\\appium-desktop\\resources\\app\\node_modules\\appium\\node_modules\\appium-unlock\\bin\\unlock_apk-debug.apk"][info] [ADB] Getting device platform version

[debug] [ADB] Getting connected devices...

[debug] [ADB] 1 device(s) connected

//获取待测设备 Android 版本

[debug] [ADB] Running 'D:\Android\android-sdk\platform-tools\adb.exe' with args: ["-P",5037,"-s","c1aeae297d72","shell","getprop","ro.build.version.release"][debug] [ADB] Current device property 'ro.build.version.release': 5.1.1

[debug] [ADB] Getting connected devices...

[debug] [ADB] 1 device(s) connected

[debug] [ADB] Running 'D:\Android\android-sdk\platform-tools\adb.exe' with args: ["-P",5037,"-s","c1aeae297d72","shell","wm","size"][debug] [ADB] Getting connected devices...

[debug] [ADB] 1 device(s) connected

//获取机型

[debug] [ADB] Running 'D:\Android\android-sdk\platform-tools\adb.exe' with args: ["-P",5037,"-s","c1aeae297d72","shell","getprop","ro.product.model"]

[debug] [ADB] Current device property 'ro.product.model': Redmi 3

[debug] [ADB] Getting connected devices...[debug] [ADB] 1 device(s) connected

//获取厂商的名字

[debug] [ADB] Running 'D:\Android\android-sdk\platform-tools\adb.exe' with args: ["-P",5037,"-s","c1aeae297d72","shell","getprop","ro.product.manufacturer"]

[debug] [ADB] Current device property 'ro.product.manufacturer': Xiaomi

//判断要测试的App是否已经安装。如果已经安装则重置；如果没安装则重新安装

//apk放在了/data/local/tmp/目录下，进行了md5的加密

[info] [AndroidDriver] Remote apk path is /data/local/tmp/230ef0fa6a2f3b78b7498451dc0e9dc9.apk

[debug] [ADB] Getting connected devices...

[debug] [ADB] 1 device(s) connected

//通过adb shell ls 查看目录

//检查/data/local/tmp/96c0b9574b2a2af7cc0999d321055f7c.apk是否存在

[debug] [ADB] Running 'D:\Android\android-sdk\platform-tools\adb.exe' with args: ["-P",5037,"-s","c1aeae297d72","shell","ls","/data/local/tmp/230ef0fa6a2f3b78b7498451dc0e9dc9.apk"][debug] [AndroidDriver] Checking if app is installed

//检查待测apk是否安装

[debug] [ADB] Getting install status for io.selendroid.testapp

[debug] [ADB] Getting connected devices...

[debug] [ADB] 1 device(s) connected

[debug] [ADB] Running 'D:\Android\android-sdk\platform-tools\adb.exe' with args: ["-P",5037,"-s","c1aeae297d72","shell","pm","list","packages","io.selendroid.testapp"][debug] [ADB] App is installed

[info] [AndroidDriver] Apk is already on remote and installed, resetting

//使用adb shell am force-stop +包名，强制停止应用

[debug] [AndroidDriver] Running fast reset (stop and clear)

[debug] [ADB] Getting connected devices...[debug] [ADB] 1 device(s) connected

[debug] [ADB] Running 'D:\Android\android-sdk\platform-tools\adb.exe' with args: ["-P",5037,"-s","c1aeae297d72","shell","am","force-stop","io.selendroid.testapp"][debug]

```
[ADB] Getting connected devices...
[debug] [ADB] 1 device(s) connected
```

// "abd shell pm clear +包名"，清理待测 App 的数据

```
[debug] [ADB] Running 'D:\Android\android-sdk\platform-tools\adb.exe' with args:
["-P",5037,"-s","c1aeae297d72","shell","pm","clear","io.selendroid.testapp"][debug]
[AndroidDriver] Extracting strings from apk F:\selendroid-test-app-0.17.0.apk undefined
C:\Users\dh\AppData\Local\Temp\io.selendroid.testapp
```

//语言默认

```
[debug] [ADB] Extracting strings for language: default
[debug] [ADB] Device API level: 22
[debug] [ADB] Getting connected devices...
[debug] [ADB] 1 device(s) connected
```

//通过 adb shell getprop persist.sys.language 获取设备语言

```
[debug] [ADB] Running 'D:\Android\android-sdk\platform-tools\adb.exe' with args:
["-P",5037,"-s","c1aeae297d72","shell","getprop","persist.sys.language"]
[debug] [ADB] Current device property 'persist.sys.language': zh[debug] [ADB] No strings.xml
for language 'zh', getting default strings.xml[debug] [ADB] Reading strings from converted
strings.json
[debug] [ADB] Running 'D:\Android\android-sdk\platform-tools\adb.exe' with args:
["-P",5037,"-s","c1aeae297d72","push","C:\\Users\\dh\\AppData\\Local\\Temp\\io.selendro
id.testapp\\strings.json","/data/local/tmp"]
```

//端口映射，转发计算机的 4724 端口到设备的 4724 端口上，发给 Appium httpserver 的内容，经过 httpserver
//后直接发给设备，并开启设备上基于 AppiumBootstrap 的 Socket 服务

```
[debug] [AndroidBootstrap] Watching for bootstrap disconnect
[debug] [ADB] Forwarding system: 4724 to device: 4724
[debug] [ADB] Running 'D:\Android\android-sdk\platform-tools\adb.exe' with args:
["-P",5037,"-s","c1aeae297d72","forward","tcp:4724","tcp:4724"]
```

//发送 AppiumBootstrap.jar 到待测设备

```
[debug] [UiAutomator] Starting UiAutomator
[debug] [UiAutomator] Moving to state 'starting'
[debug] [UiAutomator] Parsing uiautomator jar
[debug] [UiAutomator] Found jar name: 'AppiumBootstrap.jar'
[debug] [ADB] Running 'D:\Android\android-sdk\platform-tools\adb.exe' with args:
["-P",5037,"-s","c1aeae297d72","push","C:\\Users\\dh\\AppData\\Local\\Programs\\appium-
desktop\\resources\\app\\node_modules\\appium\\node_modules\\appium-android-bootstrap\\
bootstrap\\bin\\AppiumBootstrap.jar","/data/local/tmp/"]
[debug] [ADB] Attempting to kill all uiautomator processes
[debug] [ADB] Getting all processes with uiautomator
```

[debug] [ADB] Getting connected devices...[debug] [ADB] 1 device(s) connected

[debug] [ADB] Running 'D:\Android\android-sdk\platform-tools\adb.exe' with args: ["-P",5037,"-s","c1aeae297d72","shell","ps"]

[info] [ADB] No uiautomator process found to kill, continuing...

[debug] [UiAutomator] Starting UIAutomator

//开启设备上的基于AppiumBootstrap的Socket服务

[debug] [ADB] Creating ADB subprocess with args: ["-P",5037,"-s","c1aeae297d72","shell","uiautomator","runtest","AppiumBootstrap.jar","-c","io.appium.android.bootstrap.Bootstrap","-e","pkg","io.selendroid.testapp","-e","disableAndroidWatchers",false,"-e","acceptSslCerts",false][debug] [UiAutomator] Moving to state 'online'

[debug] [AndroidBootstrap] [BOOTSTRAP LOG] [debug] Registered crash watchers.

[info] [AndroidBootstrap] Android bootstrap socket is now connected

[debug] [ADB] Getting connected devices...

[debug] [AndroidBootstrap] [BOOTSTRAP LOG] [debug] Client connected

[debug] [ADB] 1 device(s) connected

[debug] [ADB] Running 'D:\Android\android-sdk\platform-tools\adb.exe' with args: ["-P",5037,"-s","c1aeae297d72","shell","dumpsys","window"][info] [AndroidDriver] Screen already unlocked, doing nothing

[debug] [ADB] Device API level: 22

[debug] [ADB] Getting connected devices...

[debug] [ADB] 1 device(s) connected

//启动应用, sessionid是bc8415e7-b66f-49ac-a966-e25a4e2684c0

[debug] [ADB] Running 'D:\Android\android-sdk\platform-tools\adb.exe' with args: ["-P",5037,"-s","c1aeae297d72","shell","am","start","-W","-n","io.selendroid.testapp/.HomeScreenActivity","-S"][info] [Appium] New AndroidDriver session created successfully, session bc8415e7-b66f-49ac-a966-e25a4e2684c0 added to master session list

[debug] [BaseDriver] Event 'newSessionStarted' logged at 1511922075296 (10:21:15 GMT+0800 (中国标准时间))

[debug] [MJSONWP] Responding to client with driver.createSession() result: {"platform":"LINUX","webStorageEnabled":false,"takesScreenshot":true,"javascriptEnabled":true,"databaseEnabled":false,"networkConnectionEnabled":true,"locationContextEnabled":false,"warnings":{},"desired":{"app":"F:\\selendroid-test-app-0.17.0.apk","appPackage":"io.selendroid.testapp","appActivity":".HomeScreenActivity","platformVersion":"5.1.1","platformName":"Android","deviceName":"REDMI NOTE 3"},"app":"F:\\selendroid-test-app-0.17.0.apk","appPackage":"io.selendroid. testapp","appActivity":".HomeScreenActivity", "platformVersion":"5.1.1","platformName":"Android","deviceName":"c1aeae297d72","deviceUDID":"c1aeae297d72","deviceScreenSize":"720x1280","deviceModel":"Redmi 3","deviceManufacturer":"Xiaomi"}

[info] [HTTP] <-- POST /wd/hub/session 200 13746 ms - 781 [info] [HTTP] --> GET /wd/hub/session/bc8415e7-b66f-49ac-a966-e25a4e2684c0 {}

[debug] [MJSONWP] Calling AppiumDriver.getSession() with args: ["bc8415e7-b66f-49ac-a966-e25a4e2684c0"]

[debug] [MJSONWP] Responding to client with driver.getSession() result: {"platform":"LINUX","webStorageEnabled":false,"takesScreenshot":true,"javascriptEnabled":true,"databaseEnabled":false,"networkConnectionEnabled":true,"locationContextEnabled":false,"warnings":{},"desired":{"app":"F:\\selendroid-test-app-0.17.0.apk","appPackage":"io.selendroid.testapp","appActivity":".HomeScreenActivity","platformVersion":"5.1.1","platformName":"Android","deviceName":"REDMI NOTE 3"},"app":"F:\\selendroid-test-app-0.17.0.apk","appPackage":"io.selendroid.testapp","appActivity":".HomeScreenActivity","platformVersion":"5.1.1","platformName":"Android","deviceName":"c1aeae297d72","deviceUDID":"c1aeae297d72","deviceScreenSize":"720x1280","deviceModel":"Redmi 3","deviceManufacturer":"Xiaomi"}

[info] [HTTP] <-- GET /wd/hub/session/bc8415e7-b66f-49ac-a966-e25a4e2684c0 200 7 ms - 781

[info] [HTTP] --> GET /wd/hub/session/bc8415e7-b66f-49ac-a966-e25a4e2684c0 {}

[debug] [MJSONWP] Calling AppiumDriver.getSession() with args: ["bc8415e7-b66f-49ac-a966-e25a4e2684c0"]

[debug] [MJSONWP] Responding to client with driver.getSession() result: {"platform":"LINUX","webStorageEnabled":false,"takesScreenshot":true,"javascriptEnabled":true,"databaseEnabled":false,"networkConnectionEnabled":true,"locationContextEnabled":false,"warnings":{},"desired":{"app":"F:\\selendroid-test-app-0.17.0.apk","appPackage":"io.selendroid.testapp","appActivity":".HomeScreenActivity","platformVersion":"5.1.1","platformName":"Android","deviceName":"REDMI NOTE 3"},"app":"F:\\selendroid-test-app-0.17.0.apk","appPackage":"io.selendroid.testapp", "appActivity":".HomeScreenActivity","platformVersion":"5.1.1","platformName":"Android","deviceName":"c1aeae297d72","deviceUDID":"c1aeae297d72","deviceScreenSize":"720x1280","deviceModel":"Redmi 3","deviceManufacturer":"Xiaomi"}

[info] [HTTP] <-- GET /wd/hub/session/bc8415e7-b66f-49ac-a966-e25a4e2684c0 200 2 ms - 781

//脚本通过post请求传递要查找的元素，在Appium服务器端进行接收，接收后解析参数并将参数通过4724端口再次传
//递给AndroidBootstrap，AndroidBootstrap完成操作后再将结果逐层返回

[info] [HTTP] --> POST /wd/hub/session/bc8415e7-b66f-49ac-a966-e25a4e2684c0/element {"using":"id","value":"io.selendroid.testapp:id/startUserRegistration"}

[debug] [MJSONWP] Calling AppiumDriver.findElement() with args: ["id","io.selendroid.testapp:id/startUserRegistration","bc8415e7-b66f-49ac-a966-e25a4e2684c0"]

[debug] [BaseDriver] Valid locator strategies for this request: xpath, id, class name, accessibility id, -android uiautomator

[debug] [BaseDriver] Valid locator strategies for this request: xpath, id, class name, accessibility id, -android uiautomator

[debug] [BaseDriver] Waiting up to 0 ms for condition

[debug] [AndroidBootstrap] Sending command to android: {"cmd":"action","action":"find","params":{"strategy":"id","selector":"io.selendroid.testapp:id/startUserRegistration","context":"","multiple":false}}

[debug] [AndroidBootstrap] [BOOTSTRAP LOG] [debug] Got data from client: {"cmd":"action","action":"find","params":{"strategy":"id","selector":"io.selendroid.testapp:id/startUserRegistration","context":"","multiple":false}}

[debug] [AndroidBootstrap] [BOOTSTRAP LOG] [debug] Got command of type ACTION

[debug] [AndroidBootstrap] [BOOTSTRAP LOG] [debug] Got command action: find

[debug] [AndroidBootstrap] [BOOTSTRAP LOG] [debug] Finding 'io.selendroid.testapp:id/startUserRegistration' using 'ID' with the contextId: '' multiple: false

//查找 id 属性为 startUserRegistration 的控件元素

[debug] [AndroidBootstrap] [BOOTSTRAP LOG] [debug] Using: UiSelector[INSTANCE=0, RESOURCE_ID=io.selendroid.testapp:id/startUserRegistration][debug] [AndroidBootstrap] [BOOTSTRAP LOG] [debug] Returning result: {"status":0,"value":{"ELEMENT":"1"}}

[debug] [AndroidBootstrap] Received command result from bootstrap

[debug] [MJSONWP] Responding to client with driver.findElement() result: {"ELEMENT":"1"}

[info] [HTTP] <-- POST /wd/hub/session/bc8415e7-b66f-49ac-a966-e25a4e2684c0/element 200 180 ms - 87

[info] [HTTP] --> POST /wd/hub/session/bc8415e7-b66f-49ac-a966-e25a4e2684c0/element/1/click {"id":"1"}

[debug] [MJSONWP] Calling AppiumDriver.click() with args: ["1","bc8415e7-b66f-49ac-a966-e25a4e2684c0"]

//执行单击命令

[debug] [AndroidBootstrap] Sending command to android:

{"cmd":"action","action":"element:click","params":{"elementId":"1"}}

[debug] [AndroidBootstrap] [BOOTSTRAP LOG] [debug] Got data from client: {"cmd":"action","action":"element:click","params":{"elementId":"1"}}

[debug] [AndroidBootstrap] [BOOTSTRAP LOG] [debug] Got command of type ACTION

[debug] [AndroidBootstrap] [BOOTSTRAP LOG] [debug] Got command action: click[debug] [AndroidBootstrap] [BOOTSTRAP LOG] [debug] Returning result: {"status":0,"value":true}

[debug] [AndroidBootstrap] Received command result from bootstrap

[debug] [MJSONWP] Responding to client with driver.click() result: true

[info] [HTTP] <-- POST /wd/hub/session/bc8415e7-b66f-49ac-a966-e25a4e2684c0/element/1/click 200 292 ms - 76

[info] [HTTP] --> POST /wd/hub/session/bc8415e7-b66f-49ac-a966-e25a4e2684c0/element {"using":"id","value":"io.selendroid.testapp:id/inputUsername"}

[debug] [MJSONWP] Calling AppiumDriver.findElement() with args: ["id","io.selendroid.testapp:id/inputUsername","bc8415e7-b66f-49ac-a966-e25a4e2684c0"]

[debug] [BaseDriver] Valid locator strategies for this request: xpath, id, class name, accessibility id, -android uiautomator

[debug] [BaseDriver] Valid locator strategies for this request: xpath, id, class name, accessibility id, -android uiautomator

[debug] [BaseDriver] Waiting up to 0 ms for condition

[debug] [AndroidBootstrap] Sending command to android: {"cmd":"action","action":"find", "params":{"strategy":"id","selector":"io.selendroid.testapp:id/inputUsername","context"

:"","multiple":false}}

[debug] [AndroidBootstrap] [BOOTSTRAP LOG] [debug] Got data from client: {"cmd":"action", "action":"find","params":{"strategy":"id","selector":"io.selendroid.testapp:id/inputUsername","context":"","multiple":false}}

[debug] [AndroidBootstrap] [BOOTSTRAP LOG] [debug] Got command of type ACTION

[debug] [AndroidBootstrap] [BOOTSTRAP LOG] [debug] Got command action: find

[debug] [AndroidBootstrap] [BOOTSTRAP LOG] [debug] Finding 'io.selendroid.testapp:id/inputUsername' using 'ID' with the contextId: '' multiple: false

[debug] [AndroidBootstrap] [BOOTSTRAP LOG] [debug] Using: UiSelector[INSTANCE=0, RESOURCE_ID=io.selendroid.testapp:id/inputUsername][debug] [AndroidBootstrap] [BOOTSTRAP LOG] [debug] Returning result: {"status":0,"value":{"ELEMENT":"2"}}

[debug] [AndroidBootstrap] Received command result from bootstrap

[debug] [MJSONWP] Responding to client with driver.findElement() result: {"ELEMENT":"2"}

[info] [HTTP] <-- POST /wd/hub/session/bc8415e7-b66f-49ac-a966-e25a4e2684c0/element 200 576 ms - 87 [info] [HTTP] --> POST /wd/hub/session/bc8415e7-b66f-49ac-a966-e25a4e2684c0/element/2/ value {"id":"2","value":["shijie"]}

[debug] [MJSONWP] Calling AppiumDriver.setValue() with args: [["shijie"],"2","bc8415e7-b66f-49ac-a966-e25a4e2684c0"]

[debug] [AndroidBootstrap] Sending command to android: {"cmd":"action","action":"element:setText","params":{"elementId":"2","text":"shijie","replace":false}}

[debug] [AndroidBootstrap] [BOOTSTRAP LOG] [debug] Got data from client: {"cmd":"action","action":"element:setText","params":{"elementId":"2","text":"shijie","replace":false}}

[debug] [AndroidBootstrap] [BOOTSTRAP LOG] [debug] Got command of type ACTION

[debug] [AndroidBootstrap] [BOOTSTRAP LOG] [debug] Got command action: setText

[debug] [AndroidBootstrap] [BOOTSTRAP LOG] [debug] Using element passed in: 2

[debug] [AndroidBootstrap] [BOOTSTRAP LOG] [debug] Attempting to clear using UiObject.clearText().[debug] [AndroidBootstrap] [BOOTSTRAP LOG] [debug] Sending plain text to element: shijie[debug] [AndroidBootstrap] [BOOTSTRAP LOG] [debug] Returning result: {"status":0,"value":true}

[debug] [AndroidBootstrap] Received command result from bootstrap

[debug] [MJSONWP] Responding to client with driver.setValue() result: true

[info] [HTTP] <-- POST /wd/hub/session/bc8415e7-b66f-49ac-a966-e25a4e2684c0/element/2/value 200 5312 ms - 76

[info] [HTTP] --> POST /wd/hub/session/bc8415e7-b66f-49ac-a966-e25a4e2684c0/element {"using":"id","value":"io.selendroid.testapp:id/inputEmail"}

[debug] [MJSONWP] Calling AppiumDriver.findElement() with args: ["id","io.selendroid.testapp:id/inputEmail","bc8415e7-b66f-49ac-a966-e25a4e2684c0"]

[debug] [BaseDriver] Valid locator strategies for this request: xpath, id, class name, accessibility id, -android uiautomator

[debug] [BaseDriver] Valid locator strategies for this request: xpath, id, class name, accessibility id, -android uiautomator

[debug] [BaseDriver] Waiting up to 0 ms for condition

[debug] [AndroidBootstrap] Sending command to android: {"cmd":"action","action":"find","params":{"strategy":"id","selector":"io.selendroid.testapp:id/inputEmail","context":"","multiple":false}}

[debug] [AndroidBootstrap] [BOOTSTRAP LOG] [debug] Got data from client: {"cmd":"action","action":"find","params":{"strategy":"id","selector":"io.selendroid.testapp:id/inputEmail","context":"","multiple":false}}

[debug] [AndroidBootstrap] [BOOTSTRAP LOG] [debug] Got command of type ACTION

[debug] [AndroidBootstrap] [BOOTSTRAP LOG] [debug] Got command action: find

[debug] [AndroidBootstrap] [BOOTSTRAP LOG] [debug] Finding 'io.selendroid.testapp:id/inputEmail' using 'ID' with the contextId: '' multiple: false

[debug] [AndroidBootstrap] [BOOTSTRAP LOG] [debug] Using: UiSelector[INSTANCE=0, RESOURCE_ID=io.selendroid.testapp:id/inputEmail][debug] [AndroidBootstrap] [BOOTSTRAP LOG] [debug] Returning result: {"status":0,"value":{"ELEMENT":"3"}}

[debug] [AndroidBootstrap] Received command result from bootstrap

[debug] [MJSONWP] Responding to client with driver.findElement() result: {"ELEMENT":"3"}

[info] [HTTP] <-- POST /wd/hub/session/bc8415e7-b66f-49ac-a966-e25a4e2684c0/element 200 508 ms - 87

[info] [HTTP] --> POST /wd/hub/session/bc8415e7-b66f-49ac-a966-e25a4e2684c0/element/3/value {"id":"3","value":["shijie@126.com"]}

[debug] [MJSONWP] Calling AppiumDriver.setValue() with args: [["shijie@126.com"],"3","bc8415e7-b66f-49ac-a966-e25a4e2684c0"]

[debug] [AndroidBootstrap] Sending command to android: {"cmd":"action","action":"element:setText","params":{"elementId":"3","text":"shijie@126.com","replace":false}}

[debug] [AndroidBootstrap] [BOOTSTRAP LOG] [debug] Got data from client: {"cmd":"action","action":"element:setText","params":{"elementId":"3","text":"shijie@126.com","replace":false}}

[debug] [AndroidBootstrap] [BOOTSTRAP LOG] [debug] Got command of type ACTION

[debug] [AndroidBootstrap] [BOOTSTRAP LOG] [debug] Got command action: setText

[debug] [AndroidBootstrap] [BOOTSTRAP LOG] [debug] Using element passed in: 3

[debug] [AndroidBootstrap] [BOOTSTRAP LOG] [debug] Attempting to clear using UiObject.clearText().[debug] [AndroidBootstrap] [BOOTSTRAP LOG] [debug] Sending plain text to element: shijie@126.com[debug] [AndroidBootstrap] [BOOTSTRAP LOG] [debug] Returning result: {"status":0,"value":true}

[debug] [AndroidBootstrap] Received command result from bootstrap

[debug] [MJSONWP] Responding to client with driver.setValue() result: true

[info] [HTTP] <-- POST /wd/hub/session/bc8415e7-b66f-49ac-a966-e25a4e2684c0/element/3/value 200 5107 ms - 76

[info] [HTTP] --> POST /wd/hub/session/bc8415e7-b66f-49ac-a966-e25a4e2684c0/element {"using":"id","value":"io.selendroid.testapp:id/inputPassword"}

[debug] [MJSONWP] Calling AppiumDriver.findElement() with args: ["id","io.selendroid.
testapp:id/inputPassword","bc8415e7-b66f-49ac-a966-e25a4e2684c0"]

[debug] [BaseDriver] Valid locator strategies for this request: xpath, id, class name,
accessibility id, -android uiautomator

[debug] [BaseDriver] Valid locator strategies for this request: xpath, id, class name,
accessibility id, -android uiautomator

[debug] [BaseDriver] Waiting up to 0 ms for condition

[debug] [AndroidBootstrap] Sending command to android: {"cmd":"action","action":"find",
"params":{"strategy":"id","selector":"io.selendroid.testapp:id/inputPassword","context"
:"","multiple":false}}

[debug] [AndroidBootstrap] [BOOTSTRAP LOG] [debug] Got data from client: {"cmd":"action",
"action":"find","params":{"strategy":"id","selector":"io.selendroid.testapp:id/inputPas
sword","context":"","multiple":false}}

[debug] [AndroidBootstrap] [BOOTSTRAP LOG] [debug] Got command of type ACTION

[debug] [AndroidBootstrap] [BOOTSTRAP LOG] [debug] Got command action: find

[debug] [AndroidBootstrap] [BOOTSTRAP LOG] [debug] Finding 'io.selendroid.testapp:id/
inputPassword' using 'ID' with the contextId: '' multiple: false

[debug] [AndroidBootstrap] [BOOTSTRAP LOG] [debug] Using: UiSelector[INSTANCE=0,
RESOURCE_ID=io.selendroid.testapp:id/inputPassword][debug] [AndroidBootstrap] [BOOTSTRAP
LOG] [debug] Returning result: {"status":0,"value":{"ELEMENT":"4"}}

[debug] [AndroidBootstrap] Received command result from bootstrap

[debug] [MJSONWP] Responding to client with driver.findElement() result: {"ELEMENT":"4"}

[info] [HTTP] <-- POST /wd/hub/session/bc8415e7-b66f-49ac-a966-e25a4e2684c0/element 200 504
ms - 87

[info] [HTTP] --> POST /wd/hub/session/bc8415e7-b66f-49ac-a966-e25a4e2684c0/element/4/
value {"id":"4","value":["123456"]}

[debug] [MJSONWP] Calling AppiumDriver.setValue() with args: [["123456"],"4","bc8415e7-
b66f-49ac-a966-e25a4e2684c0"]

[debug] [AndroidBootstrap] Sending command to android: {"cmd":"action","action":
"element:setText","params":{"elementId":"4","text":"123456","replace":false}}

[debug] [AndroidBootstrap] [BOOTSTRAP LOG] [debug] Got data from client:
{"cmd":"action","action":"element:setText","params":{"elementId":"4","text":"123456",
"replace":false}}

[debug] [AndroidBootstrap] [BOOTSTRAP LOG] [debug] Got command of type ACTION

[debug] [AndroidBootstrap] [BOOTSTRAP LOG] [debug] Got command action: setText

[debug] [AndroidBootstrap] [BOOTSTRAP LOG] [debug] Using element passed in: 4

[debug] [AndroidBootstrap] [BOOTSTRAP LOG] [debug] Attempting to clear using
UiObject.clearText().[debug] [AndroidBootstrap] [BOOTSTRAP LOG] [debug] Sending plain text
to element: 123456[debug] [AndroidBootstrap] [BOOTSTRAP LOG] [debug] Returning result:
{"status": 0,"value":true}

[debug] [AndroidBootstrap] Received command result from bootstrap

[debug] [MJSONWP] Responding to client with driver.setValue() result: true

[info] [HTTP] <-- POST /wd/hub/session/bc8415e7-b66f-49ac-a966-e25a4e2684c0/element/4/value 200 4529 ms - 76

[info] [HTTP] --> POST /wd/hub/session/bc8415e7-b66f-49ac-a966-e25a4e2684c0/element {"using":"id","value":"io.selendroid.testapp:id/inputName"}

[debug] [MJSONWP] Calling AppiumDriver.findElement() with args: ["id","io.selendroid.testapp:id/inputName","bc8415e7-b66f-49ac-a966-e25a4e2684c0"]

[debug] [BaseDriver] Valid locator strategies for this request: xpath, id, class name, accessibility id, -android uiautomator

[debug] [BaseDriver] Valid locator strategies for this request: xpath, id, class name, accessibility id, -android uiautomator

[debug] [BaseDriver] Waiting up to 0 ms for condition

[debug] [AndroidBootstrap] Sending command to android: {"cmd":"action","action":"find","params":{"strategy":"id","selector":"io.selendroid.testapp:id/inputName","context":"","multiple":false}}

[debug] [AndroidBootstrap] [BOOTSTRAP LOG] [debug] Got data from client: {"cmd":"action","action":"find","params":{"strategy":"id","selector":"io.selendroid.testapp:id/inputName","context":"","multiple":false}}

[debug] [AndroidBootstrap] [BOOTSTRAP LOG] [debug] Got command of type ACTION

[debug] [AndroidBootstrap] [BOOTSTRAP LOG] [debug] Got command action: find

[debug] [AndroidBootstrap] [BOOTSTRAP LOG] [debug] Finding 'io.selendroid.testapp:id/inputName' using 'ID' with the contextId: '' multiple: false

[debug] [AndroidBootstrap] [BOOTSTRAP LOG] [debug] Using: UiSelector[INSTANCE=0, RESOURCE_ID=io.selendroid.testapp:id/inputName][debug] [AndroidBootstrap] [BOOTSTRAP LOG] [debug] Returning result: {"status":0,"value":{"ELEMENT":"5"}}

[debug] [AndroidBootstrap] Received command result from bootstrap

[debug] [MJSONWP] Responding to client with driver.findElement() result: {"ELEMENT":"5"}

[info] [HTTP] <-- POST /wd/hub/session/bc8415e7-b66f-49ac-a966-e25a4e2684c0/element 200 528 ms - 87

[info] [HTTP] --> POST /wd/hub/session/bc8415e7-b66f-49ac-a966-e25a4e2684c0/element/5/clear {"id":"5"}

[debug] [MJSONWP] Calling AppiumDriver.clear() with args: ["5","bc8415e7-b66f-49ac-a966-e25a4e2684c0"]

[debug] [AndroidBootstrap] Sending command to android: {"cmd":"action","action":"element:getText","params":{"elementId":"5"}}

[debug] [AndroidBootstrap] [BOOTSTRAP LOG] [debug] Got data from client: {"cmd":"action","action":"element:getText","params":{"elementId":"5"}}

[debug] [AndroidBootstrap] [BOOTSTRAP LOG] [debug] Got command of type ACTION

[debug] [AndroidBootstrap] [BOOTSTRAP LOG] [debug] Got command action: getText

[debug] [AndroidBootstrap] [BOOTSTRAP LOG] [debug] Returning result: {"status":0,"value":"Mr. Burns"}

[debug] [AndroidBootstrap] Received command result from bootstrap

[debug] [AndroidBootstrap] Sending command to android: {"cmd":"action","action":"element:click","params":{"elementId":"5"}}

[debug] [AndroidBootstrap] [BOOTSTRAP LOG] [debug] Got data from client: {"cmd":"action","action":"element:click","params":{"elementId":"5"}}

[debug] [AndroidBootstrap] [BOOTSTRAP LOG] [debug] Got command of type ACTION

[debug] [AndroidBootstrap] [BOOTSTRAP LOG] [debug] Got command action: click[debug] [AndroidBootstrap] [BOOTSTRAP LOG] [debug] Returning result: {"status":0,"value":true}

[debug] [AndroidBootstrap] Received command result from bootstrap

[debug] [AndroidDriver] Sending up to 9 clear characters to device

[debug] [AndroidDriver] Sending 9 clear characters to device

[debug] [ADB] Clearing up to 9 characters

[debug] [ADB] Getting connected devices...

[debug] [ADB] 1 device(s) connected

[debug] [ADB] Running 'D:\Android\android-sdk\platform-tools\adb.exe' with args: ["-P",5037,"-s","c1aeae297d72","shell","input","keyevent","67","112","67","112","67","112","67","112","67","112","67","112","67","112","67","112","67","112"][debug] [MJSONWP] Responding to client with driver.clear() result: null

[info] [HTTP] <-- POST /wd/hub/session/bc8415e7-b66f-49ac-a966-e25a4e2684c0/element/5/clear 200 1267 ms - 76

[info] [HTTP] --> POST /wd/hub/session/bc8415e7-b66f-49ac-a966-e25a4e2684c0/element/5/value {"id":"5","value":["daming"]}

[debug] [MJSONWP] Calling AppiumDriver.setValue() with args: [["daming"],"5","bc8415e7-b66f-49ac-a966-e25a4e2684c0"]

[debug] [AndroidBootstrap] Sending command to android: {"cmd":"action","action":"element:setText","params":{"elementId":"5","text":"daming","replace":false}}

[debug] [AndroidBootstrap] [BOOTSTRAP LOG] [debug] Got data from client: {"cmd":"action","action":"element:setText","params":{"elementId":"5","text":"daming","replace":false}}

[debug] [AndroidBootstrap] [BOOTSTRAP LOG] [debug] Got command of type ACTION

[debug] [AndroidBootstrap] [BOOTSTRAP LOG] [debug] Got command action: setText

[debug] [AndroidBootstrap] [BOOTSTRAP LOG] [debug] Using element passed in: 5[debug] [AndroidBootstrap] [BOOTSTRAP LOG] [debug] Attempting to clear using UiObject.clearText().[debug] [AndroidBootstrap] [BOOTSTRAP LOG] [debug] Sending plain text to element: daming[debug] [AndroidBootstrap] [BOOTSTRAP LOG] [debug] Returning result: {"status":0,"value":true}

[debug] [AndroidBootstrap] Received command result from bootstrap

[debug] [MJSONWP] Responding to client with driver.setValue() result: true

```
[info] [HTTP] <-- POST /wd/hub/session/bc8415e7-b66f-49ac-a966-e25a4e2684c0/element/5/
value 200 5701 ms - 76
[info] [HTTP] --> POST /wd/hub/session/bc8415e7-b66f-49ac-a966-e25a4e2684c0/element
{"using":"id","value":"io.selendroid.testapp:id/input_preferedProgrammingLanguage"}
[debug] [MJSONWP] Calling AppiumDriver.findElement() with args: ["id","io.selendroid.
testapp:id/input_preferedProgrammingLanguage","bc8415e7-b66f-49ac-a966-e25a4e2684c0"]
[debug] [BaseDriver] Valid locator strategies for this request: xpath, id, class name,
accessibility id, -android uiautomator
[debug] [BaseDriver] Valid locator strategies for this request: xpath, id, class name,
accessibility id, -android uiautomator
[debug] [BaseDriver] Waiting up to 0 ms for condition
[debug] [AndroidBootstrap] Sending command to android: {"cmd":"action","action":"find",
"params":{"strategy":"id","selector":"io.selendroid.testapp:id/input_preferedProgrammin
gLanguage","context":"","multiple":false}}
[debug] [AndroidBootstrap] [BOOTSTRAP LOG] [debug] Got data from client: {"cmd":"action",
"action":"find","params":{"strategy":"id","selector":"io.selendroid.testapp:id/input_pr
eferedProgrammingLanguage","context":"","multiple":false}}
[debug] [AndroidBootstrap] [BOOTSTRAP LOG] [debug] Got command of type ACTION
[debug] [AndroidBootstrap] [BOOTSTRAP LOG] [debug] Got command action: find
[debug] [AndroidBootstrap] [BOOTSTRAP LOG] [debug] Finding 'io.selendroid.testapp:id/input
_preferedProgrammingLanguage' using 'ID' with the contextId: '' multiple: false
[debug] [AndroidBootstrap] [BOOTSTRAP LOG] [debug] Using: UiSelector[INSTANCE=0, RESOURCE_ID
=io.selendroid.testapp:id/input_preferedProgrammingLanguage][debug] [AndroidBootstrap] [BOOTSTRAP
LOG] [debug] Returning result: {"status":0,"value":{"ELEMENT":"6"}}
[debug] [AndroidBootstrap] Received command result from bootstrap
[debug] [MJSONWP] Responding to client with driver.findElement() result: {"ELEMENT":"6"}
[info] [HTTP] <-- POST /wd/hub/session/bc8415e7-b66f-49ac-a966-e25a4e2684c0/element 200 921
ms - 87
[info] [HTTP] --> POST /wd/hub/session/bc8415e7-b66f-49ac-a966-e25a4e2684c0/element/
6/click {"id":"6"}
[debug] [MJSONWP] Calling AppiumDriver.click() with args: ["6","bc8415e7-b66f-49ac-a966-
e25a4e2684c0"]
[debug] [AndroidBootstrap] Sending command to android: {"cmd":"action","action":"element:
click","params":{"elementId":"6"}}
[debug] [AndroidBootstrap] [BOOTSTRAP LOG] [debug] Got data from client:
{"cmd":"action","action":"element:click","params":{"elementId":"6"}}
[debug] [AndroidBootstrap] [BOOTSTRAP LOG] [debug] Got command of type ACTION
[debug] [AndroidBootstrap] [BOOTSTRAP LOG] [debug] Got command action: click[debug]
[AndroidBootstrap] [BOOTSTRAP LOG] [debug] Returning result: {"status":0,"value":true}
[debug] [AndroidBootstrap] Received command result from bootstrap
```

```
[debug] [MJSONWP] Responding to client with driver.click() result: true
[info] [HTTP] <-- POST /wd/hub/session/bc8415e7-b66f-49ac-a966-e25a4e2684c0/element/6/
click 200 268 ms - 76
[info] [HTTP] --> POST /wd/hub/session/bc8415e7-b66f-49ac-a966-e25a4e2684c0/element
{"using":"-android uiautomator","value":"text(\"Scala\")"}
[debug] [MJSONWP] Calling AppiumDriver.findElement() with args: ["-android uiautomator",
"text(\"Scala\")","bc8415e7-b66f-49ac-a966-e25a4e2684c0"]
[debug] [BaseDriver] Valid locator strategies for this request: xpath, id, class name,
accessibility id, -android uiautomator
[debug] [BaseDriver] Valid locator strategies for this request: xpath, id, class name,
accessibility id, -android uiautomator
[debug] [BaseDriver] Waiting up to 0 ms for condition
[debug] [AndroidBootstrap] Sending command to android: {"cmd":"action","action":"find",
"params":{"strategy":"-android uiautomator","selector":"text(\"Scala\")","context":"","multiple":
false}}
[debug] [AndroidBootstrap] [BOOTSTRAP LOG] [debug] Got data from client: {"cmd":"action",
"action":"find","params":{"strategy":"-android uiautomator","selector":"text(\"Scala\")",
"context":"","multiple":false}}
[debug] [AndroidBootstrap] [BOOTSTRAP LOG] [debug] Got command of type ACTION
[debug] [AndroidBootstrap] [BOOTSTRAP LOG] [debug] Got command action: find
[debug] [AndroidBootstrap] [BOOTSTRAP LOG] [debug] Finding 'text("Scala")' using
'ANDROID_UIAUTOMATOR' with the contextId: '' multiple: false
[debug] [AndroidBootstrap] [BOOTSTRAP LOG] [debug] Parsing selector: text("Scala")
[debug] [AndroidBootstrap] [BOOTSTRAP LOG] [debug] UiSelector coerce type: class java.lang.
String arg: "Scala"
[debug] [AndroidBootstrap] [BOOTSTRAP LOG] [debug] Using: UiSelector[TEXT=Scala][debug]
[AndroidBootstrap] [BOOTSTRAP LOG] [debug] Returning result: {"status":0,"value":{"ELEMENT":"7"}}
[debug] [AndroidBootstrap] Received command result from bootstrap
[debug] [MJSONWP] Responding to client with driver.findElement() result: {"ELEMENT":"7"}
[info] [HTTP] <-- POST /wd/hub/session/bc8415e7-b66f-49ac-a966-e25a4e2684c0/element 200 634
ms - 87
[info] [HTTP] --> POST /wd/hub/session/bc8415e7-b66f-49ac-a966-e25a4e2684c0/element/7/
click {"id":"7"}
[debug] [MJSONWP] Calling AppiumDriver.click() with args: ["7","bc8415e7-b66f-49ac-a966-
e25a4e2684c0"]
[debug] [AndroidBootstrap] Sending command to android: {"cmd":"action","action":"element:
click","params":{"elementId":"7"}}
[debug] [AndroidBootstrap] [BOOTSTRAP LOG] [debug] Got data from client: {"cmd":"action",
"action":"element:click","params":{"elementId":"7"}}
[debug] [AndroidBootstrap] [BOOTSTRAP LOG] [debug] Got command of type ACTION
```

[debug] [AndroidBootstrap] [BOOTSTRAP LOG] [debug] Got command action: click

[debug] [AndroidBootstrap] [BOOTSTRAP LOG] [debug] Returning result: {"status":0,"value":true}

[debug] [AndroidBootstrap] Received command result from bootstrap

[debug] [MJSONWP] Responding to client with driver.click() result: true

[info] [HTTP] <-- POST /wd/hub/session/bc8415e7-b66f-49ac-a966-e25a4e2684c0/element/7/click 200 219 ms - 76

[info] [HTTP] --> POST /wd/hub/session/bc8415e7-b66f-49ac-a966-e25a4e2684c0/element {"using":"id","value":"io.selendroid.testapp:id/input_adds"}

[debug] [MJSONWP] Calling AppiumDriver.findElement() with args: ["id","io.selendroid.testapp:id/input_adds","bc8415e7-b66f-49ac-a966-e25a4e2684c0"]

[debug] [BaseDriver] Valid locator strategies for this request: xpath, id, class name, accessibility id, -android uiautomator

[debug] [BaseDriver] Valid locator strategies for this request: xpath, id, class name, accessibility id, -android uiautomator

[debug] [BaseDriver] Waiting up to 0 ms for condition

[debug] [AndroidBootstrap] Sending command to android: {"cmd":"action","action":"find","params":{"strategy":"id","selector":"io.selendroid.testapp:id/input_adds","context":"","multiple":false}}[debug] [AndroidBootstrap] [BOOTSTRAP LOG] [debug] Got data from client: {"cmd":"action","action":"find","params":{"strategy":"id","selector":"io.selendroid.testapp:id/input_adds","context":"","multiple":false}}

[debug] [AndroidBootstrap] [BOOTSTRAP LOG] [debug] Got command of type ACTION

[debug] [AndroidBootstrap] [BOOTSTRAP LOG] [debug] Got command action: find

[debug] [AndroidBootstrap] [BOOTSTRAP LOG] [debug] Finding 'io.selendroid.testapp:id/input_adds' using 'ID' with the contextId: '' multiple: false

[debug] [AndroidBootstrap] [BOOTSTRAP LOG] [debug] Using: UiSelector[INSTANCE=0, RESOURCE_ID=io.selendroid.testapp:id/input_adds][debug] [AndroidBootstrap] [BOOTSTRAP LOG] [debug] Returning result: {"status":0,"value":{"ELEMENT":"8"}}

[debug] [AndroidBootstrap] Received command result from bootstrap

[debug] [MJSONWP] Responding to client with driver.findElement() result: {"ELEMENT":"8"}

[info] [HTTP] <-- POST /wd/hub/session/bc8415e7-b66f-49ac-a966-e25a4e2684c0/element 200 636 ms - 87

[info] [HTTP] --> POST /wd/hub/session/bc8415e7-b66f-49ac-a966-e25a4e2684c0/element/8/click {"id":"8"}

[debug] [MJSONWP] Calling AppiumDriver.click() with args: ["8","bc8415e7-b66f-49ac-a966-e25a4e2684c0"]

[debug] [AndroidBootstrap] Sending command to android: {"cmd":"action","action":"element:click","params":{"elementId":"8"}}

[debug] [AndroidBootstrap] [BOOTSTRAP LOG] [debug] Got data from client: {"cmd":"action","action":"element:click","params":{"elementId":"8"}}

[debug] [AndroidBootstrap] [BOOTSTRAP LOG] [debug] Got command of type ACTION

[debug] [AndroidBootstrap] [BOOTSTRAP LOG] [debug] Got command action: click[debug] [AndroidBootstrap] [BOOTSTRAP LOG] [debug] Returning result: {"status":0,"value":true}

[debug] [AndroidBootstrap] Received command result from bootstrap

[debug] [MJSONWP] Responding to client with driver.click() result: true

[info] [HTTP] <-- POST /wd/hub/session/bc8415e7-b66f-49ac-a966-e25a4e2684c0/element/8/click 200 689 ms - 76

[info] [HTTP] --> POST /wd/hub/session/bc8415e7-b66f-49ac-a966-e25a4e2684c0/element {"using":"id","value":"io.selendroid.testapp:id/btnRegisterUser"}

[debug] [MJSONWP] Calling AppiumDriver.findElement() with args: ["id","io.selendroid.testapp:id/btnRegisterUser","bc8415e7-b66f-49ac-a966-e25a4e2684c0"]

[debug] [BaseDriver] Valid locator strategies for this request: xpath, id, class name, accessibility id, -android uiautomator

[debug] [BaseDriver] Valid locator strategies for this request: xpath, id, class name, accessibility id, -android uiautomator

[debug] [BaseDriver] Waiting up to 0 ms for condition

[debug] [AndroidBootstrap] Sending command to android: {"cmd":"action","action":"find","params":{"strategy":"id","selector":"io.selendroid.testapp:id/btnRegisterUser","context":"","multiple":false}}

[debug] [AndroidBootstrap] [BOOTSTRAP LOG] [debug] Got data from client: {"cmd":"action","action":"find","params":{"strategy":"id","selector":"io.selendroid.testapp:id/btnRegisterUser","context":"","multiple":false}}

[debug] [AndroidBootstrap] [BOOTSTRAP LOG] [debug] Got command of type ACTION

[debug] [AndroidBootstrap] [BOOTSTRAP LOG] [debug] Got command action: find

[debug] [AndroidBootstrap] [BOOTSTRAP LOG] [debug] Finding 'io.selendroid.testapp:id/btnRegisterUser' using 'ID' with the contextId: '' multiple: false

[debug] [AndroidBootstrap] [BOOTSTRAP LOG] [debug] Using: UiSelector[INSTANCE=0, RESOURCE_ID=io.selendroid.testapp:id/btnRegisterUser][debug] [AndroidBootstrap] [BOOTSTRAP LOG] [debug] Returning result: {"status":0,"value":{"ELEMENT":"9"}}

[debug] [AndroidBootstrap] Received command result from bootstrap

[debug] [MJSONWP] Responding to client with driver.findElement() result: {"ELEMENT":"9"}

[info] [HTTP] <-- POST /wd/hub/session/bc8415e7-b66f-49ac-a966-e25a4e2684c0/element 200 452 ms - 87

[info] [HTTP] --> POST /wd/hub/session/bc8415e7-b66f-49ac-a966-e25a4e2684c0/element/9/click {"id":"9"}

[debug] [MJSONWP] Calling AppiumDriver.click() with args: ["9","bc8415e7-b66f-49ac-a966-e25a4e2684c0"]

[debug] [AndroidBootstrap] Sending command to android: {"cmd":"action","action":"element:click","params":{"elementId":"9"}}

```
[debug] [AndroidBootstrap] [BOOTSTRAP LOG] [debug] Got data from client: {"cmd":"action",
"action":"element:click","params":{"elementId":"9"}}
[debug] [AndroidBootstrap] [BOOTSTRAP LOG] [debug] Got command of type ACTION
[debug] [AndroidBootstrap] [BOOTSTRAP LOG] [debug] Got command action: click[debug]
[AndroidBootstrap] [BOOTSTRAP LOG] [debug] Returning result: {"status":0,"value":true}
[debug] [AndroidBootstrap] Received command result from bootstrap
[debug] [MJSONWP] Responding to client with driver.click() result: true
[info] [HTTP] <-- POST /wd/hub/session/bc8415e7-b66f-49ac-a966-e25a4e2684c0/element/9/
click 200 372 ms - 76
[info] [HTTP] --> POST /wd/hub/session/bc8415e7-b66f-49ac-a966-e25a4e2684c0/element
{"using":"id","value":"label_name_data"}
[debug] [MJSONWP] Calling AppiumDriver.findElement() with args: ["id","label_name_data",
"bc8415e7-b66f-49ac-a966-e25a4e2684c0"]
[debug] [BaseDriver] Valid locator strategies for this request: xpath, id, class name,
accessibility id, -android uiautomator
[debug] [BaseDriver] Valid locator strategies for this request: xpath, id, class name,
accessibility id, -android uiautomator
[debug] [BaseDriver] Waiting up to 0 ms for condition
[debug] [AndroidBootstrap] Sending command to android: {"cmd":"action","action":"find",
"params":{"strategy":"id","selector":"label_name_data","context":"","multiple":false}}
[debug] [AndroidBootstrap] [BOOTSTRAP LOG] [debug] Got data from client: {"cmd":"action",
"action":"find","params":{"strategy":"id","selector":"label_name_data","context":"","mu
ltiple":false}}
[debug] [AndroidBootstrap] [BOOTSTRAP LOG] [debug] Got command of type ACTION
[debug] [AndroidBootstrap] [BOOTSTRAP LOG] [debug] Got command action: find
[debug] [AndroidBootstrap] [BOOTSTRAP LOG] [debug] Finding 'label_name_data' using 'ID' with
the contextId: '' multiple: false
[debug] [AndroidBootstrap] [BOOTSTRAP LOG] [debug] Using: UiSelector[INSTANCE=0, RESOURCE_ID
=io.selendroid.testapp:id/label_name_data][debug]   [AndroidBootstrap]   [BOOTSTRAP   LOG]
[debug] Returning result: {"status":0,"value":{"ELEMENT":"10"}}
[debug] [AndroidBootstrap] Received command result from bootstrap
[debug] [MJSONWP] Responding to client with driver.findElement() result: {"ELEMENT":"10"}
[info] [HTTP] <-- POST /wd/hub/session/bc8415e7-b66f-49ac-a966-e25a4e2684c0/element 200 531
ms - 88
[info] [HTTP] --> GET /wd/hub/session/bc8415e7-b66f-49ac-a966-e25a4e2684c0/element/10/
text {}
[debug] [MJSONWP] Calling AppiumDriver.getText() with args: ["10","bc8415e7-b66f-49ac-a966-
e25a4e2684c0"]
[debug] [AndroidBootstrap] Sending command to android: {"cmd":"action","action":"element:
getText","params":{"elementId":"10"}}
```

[debug] [AndroidBootstrap] [BOOTSTRAP LOG] [debug] Got data from client: {"cmd":"action","action":"element:getText","params":{"elementId":"10"}}

[debug] [AndroidBootstrap] [BOOTSTRAP LOG] [debug] Got command of type ACTION

[debug] [AndroidBootstrap] [BOOTSTRAP LOG] [debug] Got command action: getText

[debug] [AndroidBootstrap] [BOOTSTRAP LOG] [debug] Returning result: {"status":0,"value":"daming"}

[debug] [AndroidBootstrap] Received command result from bootstrap

[debug] [MJSONWP] Responding to client with driver.getText() result: "daming"

[info] [HTTP] <-- GET /wd/hub/session/bc8415e7-b66f-49ac-a966-e25a4e2684c0/element/10/text 200 28 ms - 80

//driver.quit 断开连接，断开前要做几件事

//1.强行终止测试 App

//2.按下 Home 键，回到主界面

//3.关闭设备上的 AndroidBootstrap Socket 服务

//4.强行终止 Unlock App 进程

[info] [HTTP] --> DELETE /wd/hub/session/bc8415e7-b66f-49ac-a966-e25a4e2684c0 {}

[debug] [MJSONWP] Calling AppiumDriver.deleteSession() with args: ["bc8415e7-b66f-49ac-a966-e25a4e2684c0"]

[debug] [BaseDriver] Event 'quitSessionRequested' logged at 1511922105004 (10:21:45 GMT+0800 (中国标准时间))

[info] [Appium] Removing session bc8415e7-b66f-49ac-a966-e25a4e2684c0 from our master session list

[debug] [AndroidDriver] Shutting down Android driver

[debug] [ADB] Getting connected devices...

[debug] [ADB] 1 device(s) connected

//使用"am force-stop+包名"强行终止测试 App

[debug] [ADB] Running 'D:\Android\android-sdk\platform-tools\adb.exe' with args: ["-P",5037,"-s","c1aeae297d72","shell","am","force-stop","io.selendroid.testapp"][debug] [ADB] Pressing the HOME button

[debug] [ADB] Getting connected devices...

[debug] [ADB] 1 device(s) connected

//按下 Home 键，回到设备主界面

[debug] [ADB] Running 'D:\Android\android-sdk\platform-tools\adb.exe' with args: ["-P",5037,"-s","c1aeae297d72","shell","input","keyevent",3][debug]   [AndroidBootstrap]   Sending command to android: {"cmd":"shutdown"}

[debug] [AndroidBootstrap] [BOOTSTRAP LOG] [debug] Got data from client: {"cmd":"shutdown"}

[debug] [AndroidBootstrap] [BOOTSTRAP LOG] [debug] Got command of type SHUTDOWN

[debug] [AndroidBootstrap] [BOOTSTRAP LOG] [debug] Returning result: {"status":0,"value": "OK, shutting down"}

[debug] [AndroidBootstrap] [BOOTSTRAP LOG] [debug] Closed client connection

[debug] [AndroidBootstrap] [UIAUTO STDOUT] INSTRUMENTATION_STATUS: numtests=1[debug] [AndroidBootstrap] [UIAUTO STDOUT] INSTRUMENTATION_STATUS: stream=.[debug] [AndroidBootstrap] [UIAUTO STDOUT] INSTRUMENTATION_STATUS: id=UiAutomatorTestRunner[debug] [AndroidBootstrap] [UIAUTO STDOUT] INSTRUMENTATION_STATUS: test=testRunServer[debug] [AndroidBootstrap] [UIAUTO STDOUT] INSTRUMENTATION_STATUS: class=io.appium.android.bootstrap. Bootstrap[debug] [AndroidBootstrap] [UIAUTO STDOUT] INSTRUMENTATION_STATUS: current=1[debug] [AndroidBootstrap] [UIAUTO STDOUT] INSTRUMENTATION_STATUS_CODE: 0[debug] [AndroidBootstrap] [UIAUTO STDOUT] INSTRUMENTATION_STATUS: stream=[debug] [AndroidBootstrap] [UIAUTO STDOUT] Test results for WatcherResultPrinter=.[debug] [AndroidBootstrap] [UIAUTO STDOUT] Time: 32.418[debug] [AndroidBootstrap] [UIAUTO STDOUT] OK (1 test)[debug] [AndroidBootstrap] [UIAUTO STDOUT] INSTRUMENTATION_STATUS_CODE: -1[debug] [AndroidBootstrap] Received command result from bootstrap

//关闭UiAutomator

[debug] [UiAutomator] Shutting down UiAutomator

[debug] [UiAutomator] Moving to state 'stopping'

[debug] [UiAutomator] UiAutomator shut down normally

[debug] [UiAutomator] Moving to state 'stopped'

[debug] [ADB] Attempting to kill all uiautomator processes

[debug] [ADB] Getting all processes with uiautomator

[debug] [ADB] Getting connected devices...

[debug] [ADB] 1 device(s) connected

[debug] [ADB] Running 'D:\Android\android-sdk\platform-tools\adb.exe' with args: ["-P",5037,"-s","claeae297d72","shell","ps"][info] [ADB] No uiautomator process found to kill, continuing...

[debug] [UiAutomator] Moving to state 'stopped'

[debug] [Logcat] Stopping logcat capture

[debug] [ADB] Getting connected devices...

[debug] [ADB] 1 device(s) connected

//关闭Unlock App进程

[debug] [ADB] Running 'D:\Android\android-sdk\platform-tools\adb.exe' with args: ["-P",5037,"-s","claeae297d72","shell","am","force-stop","io.appium.unlock"][debug] [AndroidDriver] Not cleaning generated files. Add `clearSystemFiles` capability if wanted.

[debug] [BaseDriver] Event 'quitSessionFinished' logged at 1511922107322 (10:21:47 GMT+0800 (中国标准时间))

[debug] [MJSONWP] Received response: null

[debug] [MJSONWP] But deleting session, so not returning

[debug] [MJSONWP] Responding to client with driver.deleteSession() result: null

[info] [HTTP] <-- DELETE /wd/hub/session/bc8415e7-b66f-49ac-a966-e25a4e2684c0 200 2319 ms - 76

第 8 章

Appium 数据驱动测试框架封装实战

测试框架总体而言可以参考软件开发框架来构建。下面是从软件开发框架原则中提取的测试框架的属性。

- 测试框架是测试开发过程中提取特定领域测试方法共性部分形成的体系结构。
- 测试框架的作用：重用测试设计原则和测试经验，调整部分内容便可满足需求，提高测试用例设计开发质量，降低成本，缩短时间。
- 不同测试技术领域有不同的测试框架类型。
- 测试工程师需要结合自己的测试对象知识把测试框架转化成自己的测试用例。
- 测试框架是供测试人员开发相应领域测试用例的测试分析设计工具。
- 测试框架不是测试用例集，而是通用的、具有一般性的系统主体部分。测试人员像做填空一样，根据具体业务完成特定应用系统中与众不同的特殊部分。
- 测试设计模式的思想（等价类/边界值）在测试框架中进行应用。

框架特色如下。

- 框架采用 Maven 进行构建，方便快速更新 Jar 包。
- 集成了 TestNG 的 Jar 包，使用它来进行测试时，相当于执行者通过在配置文件中指定待测试的模块，轻松实现测试。
- 集成了 Log4j，全面地保存日志，无论是成功或者失败，都会生成多角度的日志信息，配合截图功能图，更快速地定位出错的代码。
- 集成了 ReportNG，生成 HTML 和 XML 形式的报告。展现当前执行的结果信息及执行的数量和时间等信息，方便统计查看。

8.1 自动化测试规划与设计

测试框架的整体架构如图 8-1 所示。

4 个包如下。

- com.shijie.base，用于实现可复用的复杂业务逻辑封装。
- com.shijie.pages，页面控件元素识别，存放每个页面元素的 Java 类。
- com.shijie.testCase，具体的测试类。

- com.shijie.util，公共类库，实现配置文件的读取、Excel 数据驱动、页面操作重新封装等。

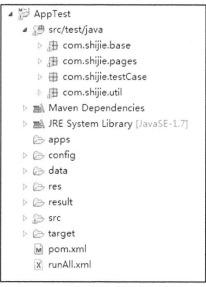

图 8-1 测试框架的整体架构

文件夹的作用如下。

- apps，存放待测 App。
- config，配置文件，包括驱动配置文件、日志配置文件以及常用路径配置文件。
- data，测试数据，存放 Excel 测试用例文件。
- res，资源文件，放置浏览器驱动。
- result，结果文件，包括 log（日志）目录、screenshot（测试截图）目录、test-report（测试报告）目录。
- pom.xml 文件，配置 Maven 依赖。
- runAll.xml 文件，配置 TestNG 运行文件。

8.2　配置 Maven 与创建 Maven 项目

5.2 节介绍了 Maven 的安装，下面将学习配置本地 Maven 仓库，并创建 Maven 项目。

8.2.1 配置 Maven

要修改 Maven 本地仓库，进入环境变量设定目录"% M2_HOME%\conf"中，修改 settings.xml 文件中的内容。修改的目的为设置存放下载的 Jar 包的本地仓库地址。建议将下载的 Jar 包存放在"% M2_HOME %/repository"指定的地址中。

1. 修改 Maven 本地仓库

首先，打开解压之后的 Maven 目录。然后，进入 conf 目录下，如图 8-2 所示。

图 8-2 conf 目录

使用编辑工具（推荐 UE 或者 Sublime Text，笔者采用 Sublime Text 打开文件）打开 settings.xml，找到 localRepository 关键字，如图 8-3 所示。去掉默认加在 localRepository 字段上的注释（<!-- -->），使该字段处于可用状态。

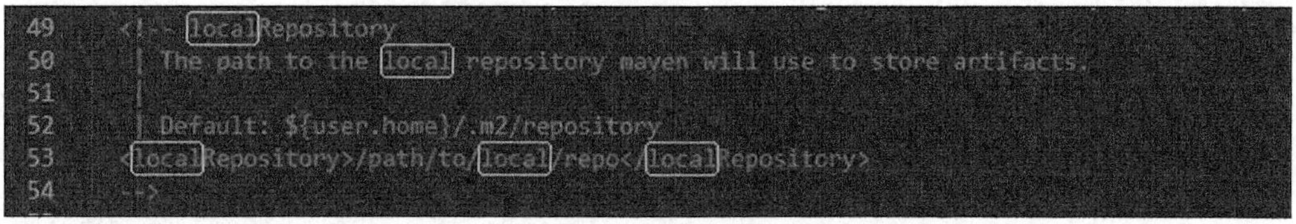

图 8-3 localRepository

在<localRepository>与</localRepository>之间配置本地仓库的地址，笔者修改为"D:/09工具/apache-maven-3.5.0-bin/apache-maven-3.5.0"，修改之后的配置如图 8-4 所示（注意，如果存放到 C 盘中，一定给该文件夹赋予权限）。

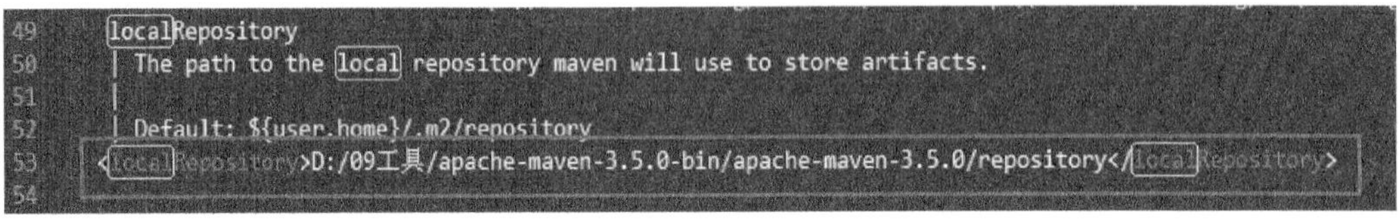

第 8 章 Appium 数据驱动测试框架封装实战

```
49  <localRepository>
50    The path to the local repository maven will use to store artifacts.
51  |
52    Default: ${user.home}/.m2/repository
53  <localRepository>D:/09工具/apache-maven-3.5.0-bin/apache-maven-3.5.0/repository</localRepository>
54
```

图 8-4　配置 Maven 本地仓库

2. 配置镜像

接下来还需要配置一个镜像（mirror），镜像就是指定从哪里下载 Jar 包，可以指定国外的，也可以指定国内的镜像地址。这里使用的是本地仓库。先找到 mirrors 这个标签，里面的英文提示在这个标签下填写镜像地址，也给出了参考格式，如图 8-5 所示。

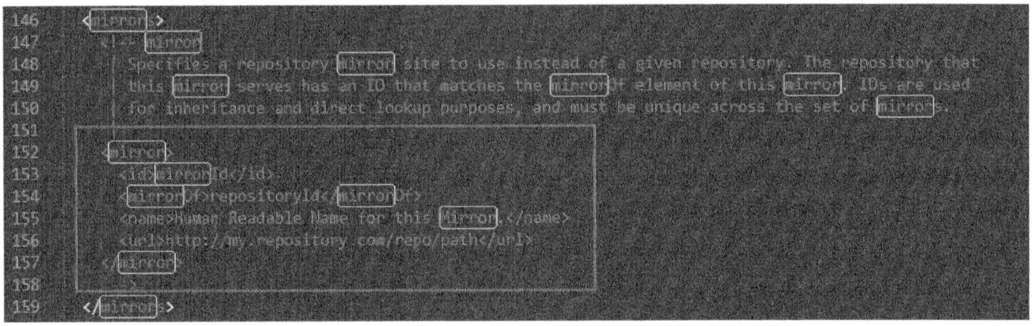

图 8-5　配置 Maven 镜像

配置本地镜像的地址，如图 8-6 所示。

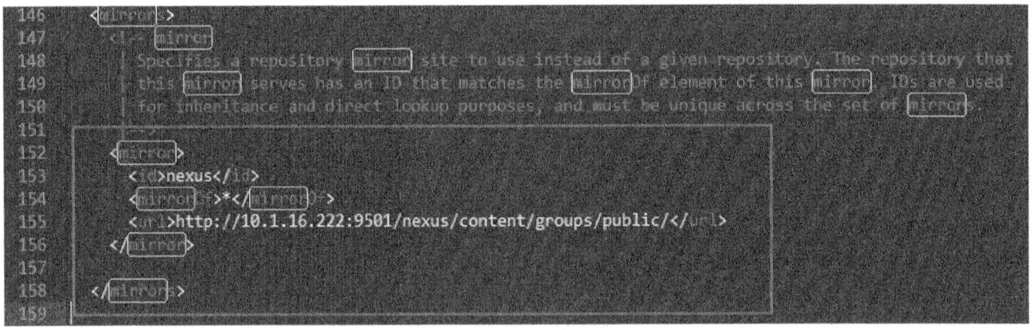

图 8-6　配置 Maven 本地镜像的地址

此时还可以增加一个或者多个稳定镜像并填写在 mirrors 标签之间。这里提供两个稳定的镜像（注意，这里的镜像地址都是远程地址，或者使用本地仓库的地址），直接粘贴到 mirrors 标签里面即可，如图 8-7 所示。

```xml
<mirror>
<!-- 
  Specifies a repository mirror site to use instead of a given repository. The repository that
  this mirror serves has an ID that matches the mirrorOf element of this mirror. IDs are used
  for inheritance and direct lookup purposes, and must be unique across the set of mirrors.
-->
<mirror>
  <id>nexus</id>
  <mirrorOf>*</mirrorOf>
  <url>http://10.1.16.222:9501/nexus/content/groups/public/</url>
</mirror>

<mirror>
  <id>aliyun-nexus</id>
  <mirrorOf>central</mirrorOf>
  <name>Nexus</name>
  <url>http://maven.aliyun.com/nexus/content/groups/public/</url>
</mirror>

<mirror>
  <id>repo2</id>
  <mirrorOf>central</mirrorOf>
  <name>repo2center</name>
  <url>http://repo2.maven.org/maven2/</url>
</mirror>
</mirror>
```

图 8-7　配置稳定镜像

刚才修改 Maven 全局配置文件，也就是说，如果没有设置用户自定义配置文件，默认会加载全局配置文件。如果设置了用户自定义配置文件，就会加载用户自定义配置文件。用户自定义配置文件，在 Maven 中是可以自己选择存放路径的。

3. 在 Eclipse 中配置 Maven

新版的 Eclipse 已经集成了新的 Maven 插件，无须额外安装。如果没有使用新的 Eclipse，可能没有集成 Maven 插件，用户需要下载 M2Eclipse 插件，安装方法请参见 5.2.5 节。安装好 Maven 插件后，开始在 Eclipse 中配置 Maven。

打开 Eclipse，选择菜单栏中的 Windows→Preferences，在弹出的 Preferences 窗口（见图 8-8）中，在左侧面板中选择 Maven→Installations→Maven。

接下来，单击 Add 按钮，选择解压之后的 Maven 目录。目录加载进来后，在 Select the installation used to launch Maven 选项组中，勾选待选择的 Maven 目录。选好之后单击 OK 按钮，完成 Maven 插件的选中。

在 Global settings from installation directory 选项下，填写刚才在 Maven 目录下配置的 conf/settings.xml 文件路径，这个会根据填写的配置文件自动获取。

切换到 User Settings 选项，如图 8-9 所示。

图 8-8　在 Eclipse 中配置 Maven

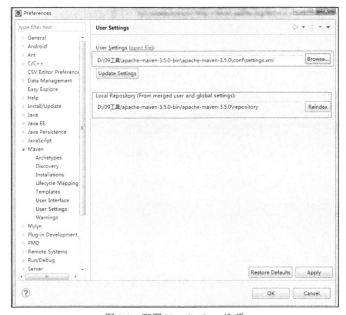

图 8-9　配置 User Settings 选项

在 User Settings 选项下，设置用户自定义配置文件，笔者直接引用了全局配置文件，所以同样填写 conf/settings.xml 文件路径，最终会以 User Settings 填写的文件为主。

在 Local Repository 选项下，设置在本地存储 Jar 包的路径，这个会根据填写的用户自定义配置文件自动获取路径。

8.2.2 创建 Maven 项目

创建 Maven 项目 AppTest。创建方式参考 5.2.3 节，创建后的结果如图 8-10 所示。

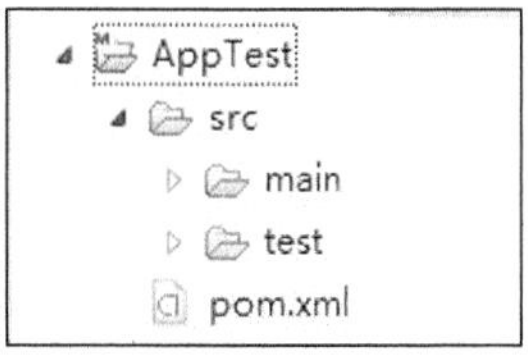

8.2.3 Maven 项目依赖包

图 8-10 创建 Maven 项目 AppTest

在 Java 项目中管理项目依赖包有两种方式。一种是直接依赖，将 Jar 包下载到本地，通过手工引入 Java 项目中；另一种是通过 Maven 进行管理，本节采用第二种方式。

参考 5.2.4 节在 Maven 项目中创建的 pom.xml 文件，直接添加依赖。

Maven 是个项目管理工具。如果不告诉它代码要使用什么 JDK 版本编译，它就会用默认的 JDK 版本来进行处理，这样就容易出现版本不匹配的问题，可能导致编译不通过。如果代码中使用了 JDK1.7 的新特性，但是 Maven 在编译的时候使用的是 JDK1.6 的版本，那么这一段代码是完全不可能编译成 .class 文件的。为了处理这一种情况，在构建 Maven 项目的时候，第一步就是配置 maven-compiler-plugin 插件。

打开 pom.xml 文件，然后切换至 pom.xml 选项卡。

复制在 Maven 仓库中查找到的 maven-compiler-plugin 包依赖代码，粘贴在<dependency>与</dependency>之间，如图 8-11 所示。

```
<dependency>
    <groupId>org.apache.maven.plugins</groupId>
    <artifactId>maven-compiler-plugin</artifactId>
    <version>3.7.0</version>
</dependency>
```

图 8-11 maven-compiler-plugin 包依赖设置

保存之后，我们发现 Maven 自动开始下载对应版本的 Jar 包。当下载完成之后，在 Maven Dependencies 目录下，单击左侧小三角，发现里面已经有相关 Jar 包。

8.3 配置 Git

Git 是一款免费、开源的分布式版本控制系统，无论项目大小，它都能高效、敏捷地处理。

Eclipse 中自带有 Git 插件。要配置 Git，在 Eclipse 的菜单栏，选择 Window→Preferences，打开 Preferences 窗口（如图 8-12 所示）。在左侧面板中，搜索 git（选择 Team→Git→Congiguation，单击 Add Entry 按钮）。然后分别建立 Key 为 user.email 和 user.name 的变量，对应的 user.email 中填写申请的 GitHub 的 E-mail，user.name 的值可以随便填写（用于区分身份）。

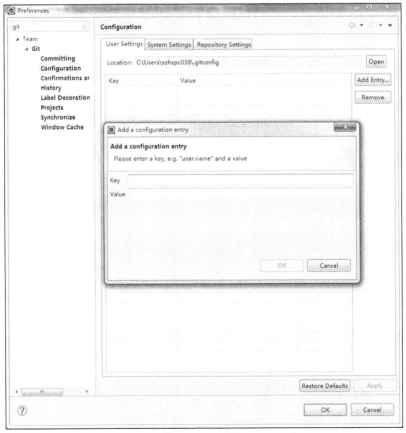

图 8-12　Git 配置

8.4 配置 SVN

SVN 是 Subversion 的简称，是一个开放源代码的版本控制系统。简单来说，SVN 用于多人共同开发同一个项目时以达到共用资源的目的。配置 SVN 的步骤如下。

（1）要安装 SVN 插件，在 Eclipse 的菜单栏中选择 Help→Eclipse Marketplace→find，输入"svn"，查找 Subversive，单击 Install 按钮安装即可，如图 8-13 所示。

图 8-13　安装 SVN

（2）SVN 无须配置，直接使用即可。

8.5 TestNG 工具

使用 TestNG 工具编写一个测试过程一般有以下 3 个步骤。

（1）将测试信息添加到测试执行文件 testng.xml 中，本测试框架在 Maven 的配置中指明了测试执行文件为 runAll.xml。

（2）编写测试逻辑并在代码中插入 TestNG 注释。

（3）运行 TestNG。

在编写测试代码之前，首先配置 TestNG 的依赖包。在 mvnrepository 网站中，查找 TestNG 的 Maven 依赖，复制到 pom.xml 中，如图 8-14 所示。

```xml
<dependencies>
    <!-- https://mvnrepository.com/artifact/org.testng/testng -->
    <dependency>
        <groupId>org.testng</groupId>
        <artifactId>testng</artifactId>
        <version>6.11</version>
        <scope>test</scope>
    </dependency>

    <dependency>
        <groupId>junit</groupId>
        <artifactId>junit</artifactId>
        <version>3.8.1</version>
        <scope>test</scope>
    </dependency>
</dependencies>
</project>
```

图 8-14 TestNG 依赖配置

复制完成后保存，在 Maven Dependencies 中可看到 TestNG 相关的依赖包已下载，如图 8-15 所示。

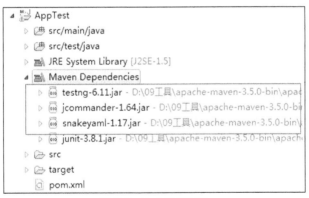

图 8-15 TestNG 依赖包

8.6 配置 Appium

8.6.1 在 Maven 中导入 Appium

当配置 Appium 包时，需要的 3 个包分别是 java-client、selenium-java、selenium-server-standalone。本节介绍如何在 Maven 中配置这 3 个依赖包。

配置后的 pom.xml 如代码清单 8-1 所示。

代码清单 8-1　　　　　　　　　配置后的 pom.xml

```
<dependency>
        <groupId>org.seleniumhq.selenium</groupId>
        <artifactId>selenium-java</artifactId>
        <version>LATEST</version>
        <!--scope>test</scope-->
</dependency>
<dependency>
        <groupId>io.appium</groupId>
        <artifactId>java-client</artifactId>
        <version>5.0.0-BETA9</version>
</dependency>
<dependency>
        <groupId>org.seleniumhq.selenium</groupId>
        <artifactId>selenium-server-standalone</artifactId>
        <version>3.4.0</version>
</dependency>
```

因为测试中可能涉及 Web App 的自动化测试，所以建议配置 selenium-api 依赖，进一步配置后的 pom.xml 如代码清单 8-2 所示。

代码清单 8-2　　　　　　　　　进一步配置后的 pom.xml

```
<dependency>
        <groupId>org.seleniumhq.selenium</groupId>
        <artifactId>selenium-api</artifactId>
        <version>LATEST</version>
        <!--scope>test</scope-->
</dependency>
```

8.6.2　创建测试脚本

下面为被测功能的相关页面描述。

(1)打开待测 App 的 HomePage,如图 8-16 所示。

(2)单击文件夹按钮,进入 RegisterPage,如图 8-17 所示。

(3)填写注册信息,进入 RegisterVerifyPage,如图 8-18 所示。

(4)单击 Register User 按钮,返回 HomePage。

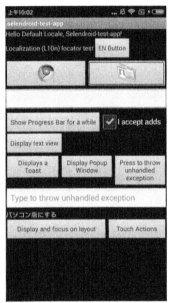
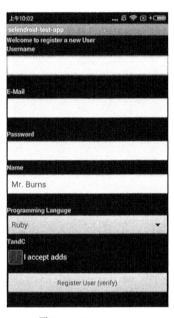

图 8-16　HomePage　　　　　图 8-17　RegisterPage　　　　　图 8-18　RegisterVerifyPage

在 Maven 项目的 AppTest 下面创建 1 个 Package,命名为 com.shijie.testScripts。在该 Package 下面创建测试类 TestRegister,该类用来完成整个测试用例。程序结构如下。

```
AppTest/
|----src/test/java
     |----com.shijie.testScripts/
          |----TestRegister.java
|----pom.xml
```

TestRegister.java 的代码如代码清单 8-3 所示。

代码清单 8-3　　　　　　　　　　TestRegister.java 的代码

```
package com.shijie.testScripts;
```

```java
import io.appium.java_client.android.AndroidDriver;
import java.io.File;
import java.net.URL;
import java.util.concurrent.TimeUnit;
import org.openqa.selenium.WebElement;
import org.openqa.selenium.remote.DesiredCapabilities;
import org.testng.Assert;
import org.testng.annotations.AfterMethod;
import org.testng.annotations.BeforeMethod;
import org.testng.annotations.Test;
import io.appium.java_client.remote.AndroidMobileCapabilityType;
import io.appium.java_client.remote.AutomationName;
import io.appium.java_client.remote.MobileCapabilityType;
public class TestRegister{
    AndroidDriver<AndroidElement> driver;

    @BeforeMethod
    public void setUp() throws Exception {
        DesiredCapabilities capabilities = new DesiredCapabilities();
        File appDir = new File("F:\\");
        File app = new File(appDir, "selendroid-test-app-0.17.0.apk");
        capabilities.setCapability("deviceName", "Redmi 3");
        capabilities.setCapability("app", app.getAbsolutePath());
        capabilities.setCapability("automationName", AutomationName.APPIUM);
        capabilities.setCapability("appPackage", "io.selendroid.testapp");
        capabilities.setCapability("appActivity",".HomeScreenActivity");

        driver = new AndroidDriver<>(new URL("http://127.0.0.1:4723/wd/hub"), capabilities);
        driver.manage().timeouts().implicitlyWait(30, TimeUnit.SECONDS);
        driver = new AndroidDriver<>(new URL("http://127.0.0.1:4723/wd/hub"), capabilities);
    }

    @Test
    public void testWebApp(){
```

```java
//对象识别
WebElement startRegister_btn = driver.findElementById("io.selendroid.testapp:id/startUserRegistration");
//单击文件夹控件,页面跳转
startRegister_btn.click();
//对象:用户名
WebElement username_txt = driver.findElementById("io.selendroid.testapp:id/inputUsername");
//输入用户名
username_txt.sendKeys("shijie");
//对象:E-mail
WebElement email_txt = driver.findElementById("io.selendroid.testapp:id/inputEmail");
//输入Email
email_txt.sendKeys("shijie@126.com");
//对象:密码
WebElement password_txt = driver.findElementById("io.selendroid.testapp:id/inputPassword");
//输入密码
password_txt.sendKeys("123456");
//对象:姓名
WebElement name_txt = driver.findElementById("io.selendroid.testapp:id/inputName");
//清空姓名默认值
name_txt.clear();
//输入姓名
name_txt.sendKeys("daming");
//对象:编程语言
WebElement languge_sel = driver.findElementById("io.selendroid.testapp:id/input_preferedProgrammingLanguage");
//进入编程语言选择界面
languge_sel.click();
//对象Scala语言
WebElement prgLanguge = driver.findElementByAndroidUIAutomator("text(\"Scala\")");
//选择Scala语言
prgLanguge.click();
//对象:接受条件
```

```
    WebElement accept_check = driver.findElementById("io.selendroid.testapp:id/
    input_adds");
    //单击接受条件
    accept_check.click();
    //对象:注册用户
    WebElement register_btn = driver.findElementById("io.selendroid.testapp:id/
    btnRegisterUser");
    //单击注册用户按钮
    register_btn.click();
    //对象:姓名确认
    WebElement label_name_data = driver.findElementById("label_name_data");
    //断言判断,姓名确认对象的值是否为前面输入的"daming"
    Assert.assertEquals(label_name_data.getText().toString(), "daming");

}




@AfterMethod
public void tearDown()
{
    driver.quit();
}

}
```

从上面的脚本中可以看出以下几个问题。

(1)测试页面元素定位和测试执行代码在同一个类中。测试页面的控件元素发生了改变,整个测试脚本有可能都要发生改变。

(2)测试数据未单独提取出来,每次测试执行后,需要手动更改脚本中的数据,以进行下一步的操作。

(3)日志信息不详细,无截图信息,无法根据错误日志定位问题。

(4)因为网络以及页面编写模式,导致页面中每个控件元素加载时间不同,需要对定位方式进行封装,以减少因为超时导致的用例失败。

下面的章节将采用各种方法修改测试类,以减少这些问题。

8.7 设计模式

除了编写自动化测试脚本的困难以外,维护脚本更是一个巨大的挑战,尤其是基于 UI 的自动化测试,界面上稍微出现变化时,就需要直接修改测试脚本,并投入大量的人力物力。长此以往,这会导致自动化测试无法深入实施。所以在编写自动化测试的过程中,重用代码能提取出来且尽量简化是自动化框架编写的一项重要工作。

8.7.1 PO 模式

Page Object(PO)模式是 Selenium 中的一种测试设计模式,Appium 也继承了这种设计模式。主要是将每一个页面设计为一个 Class,其中包含页面中需要测试的元素(按钮、输入框、标题等),这样在 Appium 测试页面中可以通过调用页面类来获取页面元素。当页面元素 id 等变化时,只需要更改测试页 Class 中页面的属性即可。这样巧妙地避免了当页面元素的 id 或者位置等定位属性变化时,需要修改整个测试代码的情况。

当然,对于 App 的 UI 测试,可能一个页面可以操作和用到的元素很少,封装成多个 Class 显得相当麻烦。对于 VerifyRegisterpage(见图 8-18),如果这个页面只与另一个页面(RegisterPage,见图 8-17)相关联,可以将相关的页面封装在一个 Class 中。对于 HomePage(见图 8-16),虽然页面元素比较少,但是作为入口页面,涉及的测试用例较多,与之相关的页面也比较多,建议还是封装成独立的页面类,这样其他页面都可以调用。

首先分析上一节的 TestRegister 类。

创建包,将其命名为 com.shijie.pages,用于存放页面对象,这个测试用例涉及 3 个页面。

- HomePage,其中包括该注册页面,以及进入注册页面的按钮。
- RegisterPage,其中包括用户名、邮箱、密码、姓名、编程语言、是否确认添加、注册按钮。
- RegisterVerifyPage,它是注册确认页面,程序结构如下。

```
AppTest/
|----src/main/java
     |----com.shijie.pages/
          |----HomePage.java
          |----RegisterPage.java
          |----RegisterVerifyPage.java
|----src/test/java
     |----com.shijie.testScripts/
          |----TestRegister.java
|----pom.xml
```

（1）创建 HomePage 类，代码如代码清单 8-4 所示。

代码清单 8-4　　　　　　　　　HomePage 类的代码

```java
package com.shijie.pages;

import io.appium.java_client.android.AndroidDriver;
import org.openqa.selenium.By;
public class HomePage {
  AndroidDriver<?> driver;
  public HomePage(AndroidDriver<?> driver){
    this.driver = driver;
  }
  //页面对象识别，进入注册页面
  private static By startRegister_btn = By.id("startUserRegistration");
  //单击"注册页面"按钮，进入注册页面，返回注册页面对象
  public RegisterPage navigate_register_page(){
    driver.findElement(startRegister_btn).click();
    return new RegisterPage(driver);
  }
}
```

在上例中，为每个页面单独定义了一个类，该类中实现了页面对对象定位的操作的内部逻辑结构。一个 Page 对象不需要关注所有细节，只关心需要的对象和操作，需要时再补充。

注意的是,如果一个页面中相同的操作产生不同的结果,则要定义多个方法,如登录操作、登录成功和登录失败需要单独定义方法。

在页面操作中,当需要页面跳转时,在方法中返回跳转后的 Page,甚至同一页面也可以返回 Page 做链式操作。

(2)创建 RegisterPage 类,代码如代码清单 8-5 所示。

代码清单 8-5　　　　　　　　　RegisterPage 类的代码

```
kage com.shijie.pages;

import org.openqa.selenium.By;
import io.appium.java_client.android.AndroidDriver;

public class RegisterPage {
  AndroidDriver<?> driver;
  public RegisterPage(AndroidDriver<?> driver) {
    // TODO Auto-generated constructor stub
    this.driver = driver;
  }
  //页面对象识别
  //用户名
  public static By username_txt = By.id("inputUsername");
  //邮箱
  public static By email_txt = By.id("inputEmail");
  //密码
  public static By password_txt = By.id("inputPassword");
  //姓名
  public static By name_txt = By.id("inputName");
  //编程语言控件
  public static By languge_sel = By.id("input_preferedProgrammingLanguage");
  //语言
  public static By prgLanguge = By.name("Scala");
  //是否确认注册
```

```
    public static By accept_check = By.id("input_adds");
    //注册按钮
    public static By register_btn = By.id("btnRegisterUser");

    public RegisterVerifyPage register_sucess(String username,String email,String password,String name){
      //页面操作，注册成功后返回注册验证页面
      driver.findElement(username_txt).sendKeys(username);
      driver.findElement(email_txt).sendKeys(email);
      driver.findElement(password_txt).sendKeys(password);
      driver.findElement(name_txt).clear();
      driver.findElement(languge_sel).sendKeys(name);
      driver.findElement(prgLanguge).click();
      driver.findElement(accept_check).click();
      driver.findElement(register_btn).click();
      return new RegisterVerifyPage(driver);
    }
}
```

（3）创建 RegisterVerifyPage 类，这个页面需要验证注册结果是否正确。需要注意的是，所有的断言操作都要放在测试代码中，而不是放在这种页面类中。RegisterVerifyPage 类的代码如代码清单 8-6 所示。

代码清单 8-6　　　　　　　　　RegisterVerifyPage 类的代码

```
package com.shijie.pages;

import org.openqa.selenium.By;
import io.appium.java_client.android.AndroidDriver;

public class RegisterVerifyPage {
  AndroidDriver<?> driver;

  public RegisterVerifyPage(AndroidDriver<?> driver) {
    this.driver = driver;
  }
```

```
//页面对象识别
public static By label_name_data = By.id("label_name_data");

public String get_name_value(){
  //注册验证页面，返回验证值
  return driver.findElement(label_name_data).getText().toString();
  }
}
```

（4）修改 TestRegister 类，通过调用页面类操作方法的形式，直接获取页面对象进行操作，实现了页面元素的定位和测试代码的分离，以降低页面元素的修改对代码的影响。当 HomePage 发生变化的时候，只需要修改 HomePage 类就可以实现，如代码清单 8-7 所示。

代码清单 8-7　　　　　　　　　　HomePage 类修改后的代码

```java
package com.shijie.testScripts;
import java.net.URL;
import io.appium.java_client.android.AndroidDriver;
import io.appium.java_client.android.AndroidElement;
import io.appium.java_client.remote.AndroidMobileCapabilityType;
import io.appium.java_client.remote.MobileCapabilityType;
import org.junit.Assert;
import org.openqa.selenium.remote.DesiredCapabilities;
import org.testng.annotations.BeforeMethod;
import org.testng.annotations.AfterMethod;
import org.testng.annotations.Test;
import com.shijie.pages.HomePage;
import com.shijie.pages.RegisterPage;
import com.shijie.pages.RegisterVerifyPage;

public class TestRegister{
  AndroidDriver<AndroidElement> driver;

  @BeforeMethod
```

```java
public void setUp() throws Exception  {
    // TODO Auto-generated method stub
    DesiredCapabilities capabilities = new DesiredCapabilities();
    capabilities.setCapability("deviceName", "Redmi 3");
    capabilities.setCapability("udid", "c1aeae297d72");
    capabilities.setCapability("platformVersion", "5.1.1");
    capabilities.setCapability("platformName", "Android");
    capabilities.setCapability("appPackage", "io.selendroid.testapp");
    capabilities.setCapability("appActivity", ".HomeScreenActivity");
    capabilities.setCapability("unicodeKeyboard", "True");
    capabilities.setCapability("noSign", "True");
    driver = new AndroidDriver<>(new URL("http://127.0.0.1:4723/wd/hub"), capabilities);
}

@Test
public void test_Register_sucess(){
    HomePage homepage = new HomePage(driver);
    RegisterPage registerpage = homepage.navigate_register_page();
    RegisterVerifyPage registerverifypage = registerpage.register_sucess("shijie",
    "shijie@126.com","123456","daming");
    Assert.assertEquals(registerverifypage.get_name_value(), "daming");
}

@AfterMethod
public void tearDown()
{
    driver.quit();
}
```

8.7.2　PageFactory 模式

PageFactory 的概念和 Page Object 应该类似，属于一种设计模式，是 Page Object 的扩展，通过注解的方式定位元素对象，需要在构造函数里面调用 PageFactory.initElements(driver, this)来初始化 PO 对象。本节介绍通过如何 PageFactory 管理对象。按照 PageFactory 设计模式，重新修改页面类。

（1）修改 HomePage 类，代码如代码清单 8-8 所示。

代码清单 8-8　　　　　　　　　HomePage 类修改后的代码

```java
package com.shijie.pages;

import org.openqa.selenium.support.PageFactory;
import io.appium.java_client.android.AndroidDriver;
import io.appium.java_client.pagefactory.AppiumFieldDecorator;
import io.appium.java_client.pagefactory.AndroidFindBy;
import io.appium.java_client.pagefactory.iOSFindBy;
import io.appium.java_client.MobileElement;

public class HomePage {
  AndroidDriver<?> driver;

  public HomePage(AndroidDriver<?> driver){
    this.driver = driver;
    PageFactory.initElements(new AppiumFieldDecorator(driver), this);
  }

  //页面对象识别，进入注册页面
  @AndroidFindBy(id="startUserRegistration")
  @iOSFindBy(id="startUserRegistration")
  public MobileElement startRegister_btn;


  //单击注册页面按钮，进入注册页面，返回注册页面对象
  public RegisterPage navigate_register_page(){
    this.startRegister_btn.click();
    return new RegisterPage(driver);
  }

}
```

（2）修改 RegisterPage 类，代码如代码清单 8-9 所示。

代码清单 8-9　　RegisterPage 类修改后的代码

```java
package com.shijie.pages;

import java.util.List;

import org.openqa.selenium.support.PageFactory;
import io.appium.java_client.android.AndroidDriver;
import io.appium.java_client.pagefactory.AppiumFieldDecorator;
import io.appium.java_client.pagefactory.AndroidFindBy;
import io.appium.java_client.pagefactory.iOSFindBy;
import io.appium.java_client.MobileElement;

public class RegisterPage {
    AndroidDriver<?> driver;
    public RegisterPage(AndroidDriver<?> driver) {
        this.driver = driver;
        PageFactory.initElements(new AppiumFieldDecorator(driver), this);
    }
    //页面对象识别
    //用户名
    @AndroidFindBy(id="inputUsername")
    @iOSFindBy(id="inputUsername")
    public MobileElement username_txt;
    //邮箱
    @AndroidFindBy(id="inputEmail")
    @iOSFindBy(id="inputEmail")
    public MobileElement email_txt;
    //密码
    @AndroidFindBy(id="inputPassword")
    @iOSFindBy(id="inputPassword")
    public MobileElement password_txt;
    //姓名
    @AndroidFindBy(id="inputName")
```

```java
@iOSFindBy(id="inputName")
public MobileElement name_txt;
//编程语言控件
@AndroidFindBy(id="input_preferedProgrammingLanguage")
@iOSFindBy(id="input_preferedProgrammingLanguage")
public MobileElement language_sel;
public List<MobileElement> prgLanguage;
//是否确认注册
@AndroidFindBy(id="input_adds")
@iOSFindBy(id="input_adds")
public MobileElement accept_check;
//注册按钮
@AndroidFindBy(id="btnRegisterUser")
@iOSFindBy(id="btnRegisterUser")
public MobileElement register_btn;


public RegisterVerifyPage register_sucess(String username,String email,String password,String name, String language,String check) {
    //页面操作,注册成功后,返回注册验证页面
    this.username_txt.sendKeys(username);
    this.email_txt.sendKeys(email);
    this.password_txt.sendKeys(password);
    this.name_txt.clear();
    this.name_txt.sendKeys(name);
    this.languge_sel.click();
    //前面的脚本中固定了待选择的语言,这里对数据进行了提取,根据参数选择不同的语言
    checkedTextView(language);
    //判断按钮是否处于已选中状态,如果未选中则执行单击操作
    if(check.equals("Yes")){
      if(!this.accept_check.isSelected())
        this.accept_check.click();}
    this.register_btn.click();
    return new RegisterVerifyPage(driver);
}
```

```java
public void checkedTextView(String language){
    //使用class属性选择所有的单选按钮,并存放在一个list中
    @SuppressWarnings("unchecked")
    List<MobileElement> checkedTextViews = (List<MobileElement>) driver.
    findElementsByClassName("android.widget.CheckedTextView");
    //使用for循环将list中的每个单选按钮进行遍历,查到name值为"Ruby"的单选按钮,
    //如果该单选按钮未处于选中状态,则调用click方法进行选择
    for(MobileElement checkedTextView:checkedTextViews){
        if(checkedTextView.getAttribute("name").equals(language)){
            if(!checkedTextView.isSelected()){
                checkedTextView.click();
            }
        }
    }
}
```

(3) 修改 RegisterVerifyPage 类,代码如代码清单 8-10 所示。

代码清单 8-10　　　　RegisterVerifyPage 类修改后的代码

```java
package com.shijie.pages;

import org.openqa.selenium.support.PageFactory;
import io.appium.java_client.android.AndroidDriver;
import io.appium.java_client.pagefactory.AppiumFieldDecorator;
import io.appium.java_client.pagefactory.AndroidFindBy;
import io.appium.java_client.pagefactory.iOSFindBy;
import io.appium.java_client.MobileElement;

public class RegisterVerifyPage {
    AndroidDriver<?> driver;

    public RegisterVerifyPage(AndroidDriver<?> driver) {
        this.driver = driver;
        PageFactory.initElements(new AppiumFieldDecorator(driver), this);
    }
```

```java
//页面对象识别
@AndroidFindBy(id="label_name_data")
@iOSFindBy(id="label_name_data")
public MobileElement label_name_data;

@AndroidFindBy(id="label_username_data")
@iOSFindBy(id="label_username_data")
public MobileElement label_username_data;

@AndroidFindBy(id="label_password_data")
@iOSFindBy(id="label_password_data")
public MobileElement label_password_data;

@AndroidFindBy(id="label_email_data")
@iOSFindBy(id="label_email_data")
public MobileElement label_email_data;

@AndroidFindBy(id="label_preferedProgrammingLanguage_data")
@iOSFindBy(id="label_preferedProgrammingLanguage_data")
public MobileElement label_preferedProgrammingLanguage_data;

@AndroidFindBy(id="label_acceptAdds_data")
@iOSFindBy(id="label_acceptAdds_data")
public MobileElement label_acceptAdds_data;
```
//增加验证项,返回验证值
```java
  public String get_name_value(){
    return this.label_name_data.getText().toString();
  }


  public String get_username_value(){
    return this.label_name_data.getText().toString();
  }


  public String get_password_value(){
```

```
    return this.label_name_data.getText().toString();
  }


  public String get_email_value(){
    return this.label_name_data.getText().toString();
  }


  public String get_preferedProgrammingLanguage_value(){
    return this.label_name_data.getText().toString();
  }


  public String get_acceptAdds_value(){
    return this.label_name_data.getText().toString();
  }
}
```

8.8 数据驱动

在 HomePage 类中未涉及测试数据，而 RegisterPage 类中涉及数据的输入，在 RegisterVerifyPage 中涉及较多的数据验证。同样，在很多的测试情况下，测试步骤不会发生变化，变化的仅仅是测试数据而已。比如在注册页面中，可能会做本地化测试，输入中文或英文。此外，还可以模拟各种合法的特殊字符串。这些测试场景下，可以使用数据驱动测试来避免重复创建雷同的测试用例。

5.3.3 节介绍了两种实现数据驱动的方式。本节再增加两种方式：一种是将测试数据写入类中，另一种是使用 CSV 的方式完成数据驱动。

首先，通过将测试数据写入类 Constants 中来实现数据驱动。

新建包，命名为 com.shijie.util，用于存放公共函数，在该包中新建 Constants 类。程序结构如下。

```
AppTest/
|----src/main/java
    |----com.shijie.pages/
        |----HomePage.java
```

```
            |----RegisterPage.java
            |----RegisterVerifyPage.java
        |----com.shijie.util/
            |----Contants.java
|----src/test/java
    |----com.shijie.testScripts/
        |----TestRegister.java
|----pom.xml
```

以 RegisterVerifyPag 类为例，将测试数据写入 Constants 类中。Constants 类的代码如代码清单 8-11 所示。

代码清单 8-11　　　　　　　　　　　Constants 类的代码

```java
package com.shijie.util;

public class Constants {
    public class 注册页面
        public static final String username = "shijie";
        public static final String eMail = "shijie@126.com";
        public static final String password = "123456";
        public static final String name = "daming";
        public static final String proLanguage = "PHP";
        public static final String accept = "true";
    }
}
```

在 TestRegister 类中导入 Constants 类，进行参数的传入，修改 TestRegister 类，如代码清单 8-12 所示。

代码清单 8-12　　　　　　　　　　　TestRegister 类

```java
package com.shijie.testScripts;

import java.net.URL;
import io.appium.java_client.android.AndroidDriver;
import io.appium.java_client.android.AndroidElement;
import com.shijie.util.Constants;
```

```java
import org.openqa.selenium.remote.DesiredCapabilities;
import org.testng.annotations.BeforeMethod;
import org.testng.annotations.AfterMethod;
import org.testng.annotations.Test;
import org.testng.Assert;
import com.shijie.pages.HomePage;
import com.shijie.pages.RegisterPage;
import com.shijie.pages.RegisterVerifyPage;
public class TestRegister{
  AndroidDriver<AndroidElement> driver;
    @BeforeClass
  public void setUp() throws Exception  {
    DesiredCapabilities capabilities = new DesiredCapabilities();
    capabilities.setCapability("deviceName", "Redmi 3");
    capabilities.setCapability("udid", "c1aeae297d72");
    capabilities.setCapability("platformVersion", "5.1.1");
    capabilities.setCapability("platformName", "Android");
    capabilities.setCapability("appPackage", "io.selendroid.testapp");
    capabilities.setCapability("appActivity", ".HomeScreenActivity");
    capabilities.setCapability("unicodeKeyboard", "True");
    capabilities.setCapability("noSign", "True");
    driver = new AndroidDriver<>(new URL("http://127.0.0.1:4723/wd/hub"), capabilities);
  }

    @Test
  public void test_Register_sucess(){
    HomePage homepage = new HomePage(driver);
    RegisterPage registerpage = homepage.navigate_register_page();
    RegisterVerifyPage registerverifypage = registerpage.register_sucess(Constants.注册页面.username,Constants.注册页面.eMail,Constants.注册页面.password,Constants.注册页面.name, Constants.注册页面.proLanguage,Constants.注册页面.accept);
    Assert.assertEquals(registerverifypage.get_name_value(), "daming");
  }

    @AfterMethod
```

```
    public void tearDown(){
      driver.quit();
    }
}
```

采用 Constants 的方法比较直观，但是如果测试数据较多，需要多次迭代，这种方法需要重写多个 Constants，因此明显不适合。Constants 适合存放测试中的常量，如用来设置数据位置等。这一节介绍采用 CSV 文件格式，通过解析 CSV 文件，来获取测试数据。

（1）配置 CSV 文件的 Maven 依赖，如代码清单 8-13 所示。

代码清单 8-13　　　　　　　　　　CSV 依赖配置

```
<dependency>
    <groupId>net.sourceforge.javacsv</groupId>
    <artifactId>javacsv</artifactId>
    <version>2.0</version>
</dependency>
```

（2）修改 Constants 类（见代码清单 8-14），用来定义数据驱动文件的位置。

代码清单 8-14　　　　　　　　　　　Constants 类

```
package com.shijie.util;

public class Constants {

    public class 注册页面{
        public static final String filepath = "F:\\book\\AppTest\\data";
        public static final String filename = "register.csv";
    }

}
```

（3）在项目根目录下创建 data 文件夹，在文件夹目录下放置数据驱动文件 register.csv。

```
no testName username eMail password name prolanguage accept verifyusername verifyeMail
1 register shijie shijie@126.com 123456 daming PHP TRUE register shijie
```

在公共函数包 com.shijie.util 中创建 DataProviderFromCsv 类，用于完成数据驱动函数的创建和数据的操作。程序结构如下。

```
AppTest/
|----src/main/java
    |----com.shijie.pages/
        |----HomePage.java
        |----RegisterPage.java
        |----RegisterVerifyPage.java
    |----com.shijie.util/
        |----Contants.java
        |----DataProviderFromCsv.java
|----src/test/java
    |----com.shijie.testScripts/
        |----TestRegister.java
|----data
        |----register.csv
|----pom.xml
```

DataProviderFromCsv 类的内容如代码清单 8-15 所示。

代码清单 8-15 　　　　　　　　DataProviderFromCsv 类

```java
package com.shijie.util;

import java.io.BufferedReader;
import java.io.FileInputStream;
import java.io.FileNotFoundException;
import java.io.FileReader;
import java.io.FileWriter;
import java.io.IOException;
import java.io.InputStreamReader;
import java.nio.charset.Charset;
import java.util.ArrayList;
import java.util.List;

import com.csvreader.CsvReader;
import com.csvreader.CsvWriter;

public class DataProviderFromCsv {
```

```java
/*作为数据驱动文件
 * */
public static Object[][] getTestData(String FileNameroot) throws IOException{
    List<Object[]> records=new ArrayList<Object[]>();
    String record;
    //设定UTF-8字符集,使用带缓冲区的字符输入流BufferedReader读取文件内容
    BufferedReader file=new BufferedReader(new InputStreamReader(new FileInputStream
    (FileNameroot),"GBK"));
    //忽略读取CSV文件的标题行(第一行)
    file.readLine();
    //遍历读取文件中除第一行外的其他所有内容并存储在名为records的ArrayList中,
    //每一行的records中存储的对象为一个String数组
    while((record=file.readLine())!=null){
      String fields[]=record.split(",");
      //System.out.println(fields[1]);
      records.add(fields);
    }
    //关闭文件对象
    file.close();
    //将存储测试数据的List转换为一个关于Object的二维数组
    Object[][] results=new Object[records.size()][];
    //设置二维数组每行的值,每行是一个Object对象
    for(int i=0;i<records.size();i++){
      results[i]=records.get(i);
    }
    return results;
  }
}
```

执行以上操作后,重新修改TestRegister类,完成从CSV的数据驱动,如代码清单8-16所示。

代码清单 8-16　　　　　　　　　　TestRegister 类

```java
package com.shijie.testScripts;

import org.testng.annotations.Test;
import org.testng.Assert;
import org.apache.log4j.Logger;
import com.shijie.base.baseActivity;
import com.shijie.pages.HomePage;
import com.shijie.pages.RegisterPage;
import com.shijie.pages.RegisterVerifyPage;



public class TestRegister{
  AndroidDriver<AndroidElement> driver;
  @BeforeClass
  public void setUp() throws Exception  {
    DesiredCapabilities capabilities = new DesiredCapabilities();
    capabilities.setCapability("deviceName", "Redmi 3");
    capabilities.setCapability("udid", "c1aeae297d72");
    capabilities.setCapability("platformVersion", "5.1.1");
    capabilities.setCapability("platformName", "Android");
    capabilities.setCapability("appPackage", "io.selendroid.testapp");
    capabilities.setCapability("appActivity", ".HomeScreenActivity");
    capabilities.setCapability("unicodeKeyboard", "True");
    capabilities.setCapability("noSign", "True");
    driver = new AndroidDriver<>(new URL("http://127.0.0.1:4723/wd/hub"), capabilities);
  }


@DataProvider(name="RegisterData" , parallel = true)
  public static Object[][] getRegisterData() throws Exception {
    return DataProviderFromCSV.getTestData(Constants.注册页面.filepath+"/"+Constants.注册页面.filename);
  }
```

```java
@Test(dataProvider = "RegisterData")
public void test_Register_sucess(String no, String testName, String username, String
eMail, String password, String name, String prolanguage, String accept,String verify
username, String verifyeMail, String verifypassword, String verifyname, String verif
yprolanguage, String verifyaccept) throws IllegalArgumentException, Exception{
HomePage homepage = null;
RegisterPage registerpage = null;
RegisterVerifyPage registerverifypage = null;
try{
  homepage = new HomePage(getDriver());
  registerpage = homepage.navigate_register_page();
  registerverifypage = registerpage.register_sucess(username, eMail, password, name,
   prolanguage, accept);
}catch(AssertionError error){
  Assert.fail("调用注册成功方法失败");
  }
try{
  Assert.assertEquals(registerverifypage.get_name_value(), name);
  }catch(AssertionError nmaeerror){
    Assert.fail("name 断言失败,查看 name 值是否正确");
  }
try{
  Assert.assertEquals(registerverifypage.get_username_value(), username);
}catch(AssertionError usernasameerror){
    Assert.fail("username 断言失败,查看 username 值是否正确");
  }
try{
  Assert.assertEquals(registerverifypage.get_password_value(), password);
}catch(AssertionError passworderror){
    Assert.fail("password 断言失败,查看 password 值是否正确");
}
try{
Assert.assertEquals(registerverifypage.get_preferedProgrammingLanguage_value(),
prolanguage);
}catch(AssertionError prolanguageerror){
    Assert.fail("preferedProgrammingLanguage 断言失败,查看 preferedProgrammingLanguage
```

```
        值是否正确");
      }
      try{
        Assert.assertEquals(registerverifypage.get_email_value(), eMail);
      }catch(AssertionError eMailerror){
        Assert.fail("email 断言失败,查看email 值是否正确");
      }
      try{
        Assert.assertEquals(registerverifypage.get_acceptAdds_value(), accept);
      }catch(AssertionError accepterror){
        Assert.fail("accept 断言失败,查看accept 值是否正确");
      }
    }
}

@AfterMethod
public void tearDown()
{
    driver.quit();
}
}
```

通过 TestRegister 脚本可以看到,DesiredCapabilities 配置是每个测试脚本都用到的,存在着代码重复。当有一个地方需要修改的时候,多个地方都要重新修改,这样不利于代码维护。下一节将讲述如何提取公用代码。

8.9 公共库

查看测试 TestRegister 脚本,从测试脚本上可以看出,对于每个测试用例,在测试执行之前,都要配置测试手机的配置信息(Desired Capabilities),在测试执行结束后要关闭测试驱动(driver.quit();)。为了提高代码的复用性,避免代码重复,这里把测试开始和结束时清理环境的代码提取出来。对于数据文件,每个测试用例都要重写一次。

创建包 com.shijie.base。在该包中创建类 BaseActivity.java,用于实现测试类执行前后的操作,以及数据提供者方法的实现。程序结构如下。

```
AppTest/
|----src/main/java
     |----com.shijie.base/
          |----BaseActivity.java
     |----com.shijie.pages/
          |----HomePage.java
          |----RegisterPage.java
          |----RegisterVerifyPage.java
     |----com.shijie.util/
          |----Contants.java
          |----DataProviderFromCsv.java
|----src/test/java
     |----com.shijie.testScripts/
          |----TestRegister.java
|----data
     |----register.csv
|----pom.xml
```

BaseActivity 类的详细内容如代码清单 8-17 所示。

代码清单 8-17　　　　　　　　　　BaseActivity 类

```java
package com.shijie.base;
import io.appium.java_client.android.AndroidDriver;
import io.appium.java_client.android.AndroidElement;
import java.net.URL;
import org.apache.log4j.PropertyConfigurator;
import org.openqa.selenium.remote.DesiredCapabilities;
import org.testng.annotations.AfterClass;
import org.testng.annotations.BeforeClass;
import org.testng.annotations.DataProvider;
import org.testng.annotations.Parameters;

import com.shijie.util.Constants;
import com.shijie.util.DataProviderFromExcel;


public class BaseActivity {
```

```java
AndroidDriver<AndroidElement> driver;

@BeforeClass
public void setUp() throws Exception {

    DesiredCapabilities capabilities = new DesiredCapabilities();
    capabilities.setCapability("deviceName", "Redmi 3");
    capabilities.setCapability("udid", "c1aeae297d72");
    capabilities.setCapability("platformVersion", "5.1.1");
    capabilities.setCapability("platformName", "Android");
    capabilities.setCapability("appPackage", "io.selendroid.testapp");
    capabilities.setCapability("appActivity", ".HomeScreenActivity");
    capabilities.setCapability("unicodeKeyboard", "True");
    capabilities.setCapability("noSign", "True");

    driver = new AndroidDriver<>(new URL("http://127.0.0.1:4723/wd/hub"), capabilities);
}


@AfterClass
/**结束测试,关闭浏览器*/
public void endTest() {
    driver.quit();
}



public AndroidDriver<AndroidElement> getDriver() {
    return driver;
}

}
```

重新修改 TestRegister 测试类,使它继承自 BaseActivity 类,如代码清单 8-18 所示。

代码清单 8-18　　　　　　　　TestRegister 类

```java
package com.shijie.testScripts;
```

```java
import java.net.URL;
import io.appium.java_client.android.AndroidDriver;
import io.appium.java_client.android.AndroidElement;
import io.appium.java_client.remote.AndroidMobileCapabilityType;
import io.appium.java_client.remote.MobileCapabilityType;
import org.openqa.selenium.remote.DesiredCapabilities;
import org.testng.annotations.BeforeMethod;
import org.testng.annotations.AfterMethod;
import org.testng.annotations.DataProvider;
import org.testng.annotations.Test;
import org.testng.Assert;
import org.apache.log4j.Logger;
import com.shijie.base.baseActivity;
import com.shijie.pages.HomePage;
import com.shijie.pages.RegisterPage;
import com.shijie.pages.RegisterVerifyPage;
import com.shijie.util.Constants;
import com.shijie.util.DataProviderFromExcel;

public class TestRegister extends BaseActivity {
    AndroidDriver<AndroidElement> driver;


    @DataProvider(name="RegisterData" , parallel = true)
    public static Object[][] getRegisterData() throws Exception {
        return DataProviderFromCsv.getTestData(Constants.注册页面.filepath+"/"+Constants.注册页面.filename);
    }


    @Test(dataProvider = " RegisterData")
    public void test_Register_sucess(String no, String testName, String username, String eMail, String password, String name, String prolanguage, String accept,String verifyusername, String verifyeMail, String verifypassword, String verifyname, String verifyprolanguage, String verifyaccept) throws IllegalArgumentException, Exception{
        HomePage homepage = null;
        RegisterPage registerpage = null;
```

```java
RegisterVerifyPage registerverifypage = null;
try{
  homepage = new HomePage(getDriver());
  registerpage = homepage.navigate_register_page();
  registerverifypage = registerpage.register_sucess(username, eMail, password, name,
  prolanguage, accept);
}catch(AssertionError error){
DataProviderFromExcel.setCellData(Integer.parseInt(no.split("\\.")[0]), DataProvider
FromExcel.getLastColumnNum(), "测试用例注册执行失败");
  Assert.fail("调用注册成功方法失败，结果写入测试用例的result列");
}
try{
Assert.assertEquals(registerverifypage.get_name_value(), verifyname);
  }catch(AssertionError nameerror){
DataProviderFromExcel.setCellData(Integer.parseInt(no.split("\\.")[0]), DataProvider
FromExcel.getLastColumnNum(), "测试用例注册执行失败");
    Assert.fail("name断言失败，查看name值是否正确");
  }
try{
  Assert.assertEquals(registerverifypage.get_username_value(), verifyusername);
}catch(AssertionError usernameerror){
DataProviderFromExcel.setCellData(Integer.parseInt(no.split("\\.")[0]), DataProvider
FromExcel.getLastColumnNum(), "测试用例注册执行失败");
    Assert.fail("username断言失败，查看username值是否正确");
  }
try{
  Assert.assertEquals(registerverifypage.get_password_value(), verifypassword);
}catch(AssertionError passworderror){
DataProviderFromExcel.setCellData(Integer.parseInt(no.split("\\.")[0]), DataProvider
FromExcel.getLastColumnNum(), "测试用例注册执行失败");
    Assert.fail("password断言失败，查看password值是否正确");
}
try{
Assert.assertEquals(registerverifypage.get_preferedProgrammingLanguage_value(),
verifyprolanguage);
}catch(AssertionError prolanguageerror){
```

```
        DataProviderFromExcel.setCellData(Integer.parseInt(no.split("\\.")[0]), DataProvid
    erFromExcel.getLastColumnNum(), "测试用例注册执行失败");
        Assert.fail("preferedProgrammingLanguage 断言失败,查看 preferedProgrammingLanguage
        值是否正确");
    }
    try{
        Assert.assertEquals(registerverifypage.get_email_value(), verifyeMail);
    }catch(AssertionError eMailerror){
    DataProviderFromExcel.setCellData(Integer.parseInt(no.split("\\.")[0]), DataProvider
    FromExcel.getLastColumnNum(), "测试用例注册执行失败");
        Assert.fail("email 断言失败,查看 email 值是否正确");
    }
    try{
        Assert.assertEquals(registerverifypage.get_acceptAdds_value(), verifyaccept);
    }catch(AssertionError accepterror){
    DataProviderFromExcel.setCellData(Integer.parseInt(no.split("\\.")[0]), DataProvider
    FromExcel.getLastColumnNum(), "测试用例注册执行失败");
        Assert.fail("accept 断言失败,查看 accept 值是否正确");
    }
    DataProviderFromExcel.setCellData(Integer.parseInt(no.split("\\.")[0]), DataProvider
    FromExcel.getLastColumnNum(), "测试用例注册执行成功");
    }
}
```

8.10 Log4j 日志

在自动化测试执行中,使用 Log4j 在日志文件中显示日志,用于监控和后续调试测试脚本。

8.10.1 在 Maven 中导入 Log4j

5.4 节介绍了如何手工配置 Log4j,本节将使用在 Maven 中管理 Jar 包的方式进行配置。修改 pom.xml,增加 Log4j 的依赖包,如代码清单 8-19 所示。

代码清单 8-19   pom.xml 的配置

```xml
<!-- https://mvnrepository.com/artifact/log4j/log4j -->
<dependency>
    <groupId>log4j</groupId>
    <artifactId>log4j</artifactId>
    <version>1.2.17</version>
</dependency>
```

8.10.2　Log4j 的使用

在前面的章节中，我们完成了数据驱动的测试脚本。在熟悉 Log4j 后，我们在测试项目中加入日志处理。

在项目目录下创建 config 文件夹。在 config 文件夹中创建 log4j.properties 文件。在测试代码中加入日志，程序结构如下。

```
AppTest/
|----src/main/java
     |----com.shijie.base/
          |----BaseActivity.java
     |----com.shijie.pages/
          |----HomePage.java
          |----RegisterPage.java
          |----RegisterVerifyPage.java
     |----com.shijie.util/
          |----Contants.java
          |----DataProviderFromCsv.java
|----src/test/java
     |----com.shijie.testScripts/
     |----TestRegister.java
|----data
     |----register.csv
|----config
     |----log4j.properties
|----pom.xml
```

步骤如下。

（1）创建 config 文件夹，并在该文件夹下创建文件 log4j.properties。配置 log4j.properties 文件，如代码清单 8-20 所示。

代码清单 8-20　　　　　　　　　log4j.properties 的配置

```
log4j.rootLogger=info, toConsole, toFile
log4j.appender.file.encoding=UTF-8
log4j.appender.toConsole=org.apache.log4j.ConsoleAppender
log4j.appender.toConsole.Target = System.out
log4j.appender.toConsole.layout=org.apache.log4j.PatternLayout
log4j.appender.toConsole.layout.ConversionPattern=[%d{yyyy-MM-dd HH:mm:ss}] [%p] %m%n
log4j.appender.toFile=org.apache.log4j.DailyRollingFileAppender
log4j.appender.toFile.file=result/log/testlog.log
log4j.appender.toFile.append=false
log4j.appender.toFile.Threshold =info
log4j.appender.toFile.layout=org.apache.log4j.PatternLayout
log4j.appender.toFile.layout.ConversionPattern=[%d{yyyy-MM-dd HH:mm:ss}] [%p] %m%n
```

（2）修改 TestRegister 类，增加日志处理，如代码清单 8-21 所示。

代码清单 8-21　　　　　　　　　TestRegister 类

```java
package com.shijie.testScripts;

import org.testng.annotations.Test;
import org.testng.Assert;
import org.apache.log4j.Logger;
import com.shijie.base.baseActivity;
import com.shijie.pages.HomePage;
import com.shijie.pages.RegisterPage;
import com.shijie.pages.RegisterVerifyPage;

public class testRegister extends baseActivity {
  Logger log = Logger.getLogger(testRegister.class.getName());


  @DataProvider(name="RegisterData" , parallel = true)
```

```java
    public static Object[][] getRegisterData() throws Exception {
        return DataProviderFromExcel.getTestData(Constants.注册页面.filepath + "." + Constants.注册页面.filename);
    }

    @Test(dataProvider = "RegisterData")
    public void test_Register_sucess(String no, String testName, String username, String eMail, String password, String name, String prolanguage, String accept,String verifyusername, String verifyeMail, String verifypassword, String verifyname, String verifyprolanguage, String verifyaccept) throws IllegalArgumentException, Exception{
        HomePage homepage = null;
        RegisterPage registerpage = null;
        RegisterVerifyPage registerverifypage = null;
        try{
            log.info("初始化home页面");
            homepage = new HomePage(getDriver());
            log.info("单击注册按钮,跳转到注册页面");
            registerpage = homepage.navigate_register_page();
            log.info("调用注册成功方法,传入参数");
            registerverifypage = registerpage.register_sucess(username, eMail, password, name, prolanguage, accept);
        }catch(AssertionError error){
            log.info("调用注册成功方法失败");
            Assert.fail("调用注册成功方法失败");
        }
        try{
            log.info("注册验证页面是否包含输入的name值");
            Assert.assertEquals(registerverifypage.get_name_value(), name);
        }catch(AssertionError nmaeerror){
            log.info("name断言失败,把结果写入测试用例的result列");
            Assert.fail("name断言失败,查看name值是否正确");
        }
        try{
            log.info("注册验证页面是否包含输入的username值");
            Assert.assertEquals(registerverifypage.get_username_value(), username);
        }catch(AssertionError usernasameerror){
```

```java
        log.info("username 断言失败,把结果写入测试用例的 result 列");
        Assert.fail("username 断言失败,查看 username 值是否正确");
    }
    try{
        log.info("注册验证页面是否包含输入的 password 值");
        Assert.assertEquals(registerverifypage.get_password_value(), password);
    }catch(AssertionError passworderror){
        log.info("password 断言失败,把结果写入测试用例的 result 列");
        Assert.fail("password 断言失败,查看 password 值是否正确");
    }
    try{
        log.info("注册验证页面是否包含输入的 preferedProgrammingLanguage 值");
        Assert.assertEquals(registerverifypage.get_preferedProgrammingLanguage_value(),
        prolanguage);
    }catch(AssertionError prolanguageerror){
        log.info("preferedProgrammingLanguage 断言失败,把结果写入测试用例的 result 列");
         Assert.fail("preferedProgrammingLanguage 断言失败,查看 preferedProgrammingLanguage
         值是否正确");
    }
    try{
        log.info("注册验证页面是否包含输入的 email 值");
        Assert.assertEquals(registerverifypage.get_email_value(), eMail);
    }catch(AssertionError eMailerror){
        log.info("email 断言失败,把结果写入测试用例的 result 列");
        Assert.fail("email 断言失败,查看 email 值是否正确");
    }
    try{
        log.info("注册验证页面是否包含输入的 accept 值");
        Assert.assertEquals(registerverifypage.get_acceptAdds_value(), accept);
    }catch(AssertionError accepterror){
        log.info("accept 断言失败,把结果写入测试用例的 result 列");
        Assert.fail("accept 断言失败,查看 accept 值是否正确");
    }
    log.info("所有断言成功,把结果写入测试用例的 result 列");
    try{
    }catch(Exception e){
```

```
        log.info("测试脚本执行成功,将结果写入测试用例失败");
    }
  }
}
```

(3) 修改 HomePage 类,如代码清单 8-22 所示。

代码清单 8-22　　　　　　　　　　HomePage 类

```java
package com.shijie.pages;

import org.apache.log4j.Logger;

public class HomePage {
  AndroidDriver<?> driver;
  Logger log = Logger.getLogger(HomePage.class.getName());
  public HomePage(AndroidDriver<?> driver){
    this.driver = driver;
    PageFactory.initElements(new AppiumFieldDecorator(driver), this);
  }

  //页面对象识别,进入注册页面
  @AndroidFindBy(id="startUserRegistration")
  @iOSFindBy(id="startUserRegistration")
  public MobileElement startRegister_btn;

  //单击注册页面按钮,进入注册页面,返回注册页面对象
  public RegisterPage navigate_register_page(){
    log.info("在 HomePage,单击进入 register 页面");
    this.startRegister_btn.click();
    return new RegisterPage(driver);
  }
}
```

修改 RegisterPage 类,如代码清单 8-23 所示。

代码清单 8-23　　　　　　　　　　RegisterPage 类

```java
package com.shijie.pages;

import java.util.List;
```

```java
import org.apache.log4j.Logger;
import org.openqa.selenium.support.PageFactory;
import com.shijie.testScripts.testRegister;
import io.appium.java_client.android.AndroidDriver;
import io.appium.java_client.pagefactory.AppiumFieldDecorator;
import io.appium.java_client.pagefactory.AndroidFindBy;
import io.appium.java_client.pagefactory.iOSFindBy;
import io.appium.java_client.MobileElement;

public class RegisterPage {
  AndroidDriver<?> driver;
  Logger log = Logger.getLogger(testRegister.class.getName());
  public RegisterPage(AndroidDriver<?> driver) {
//PropertyConfigurator.configure("F://book//AppTest//config//log4j.properties");
    this.driver = driver;
    PageFactory.initElements(new AppiumFieldDecorator(driver), this);
  }
  //页面对象识别
  //用户名
  @AndroidFindBy(id="inputUsername")
  @iOSFindBy(id="inputUsername")
  public MobileElement username_txt;
  //邮箱
  @AndroidFindBy(id="inputEmail")
  @iOSFindBy(id="inputEmail")
  public MobileElement email_txt;
  //密码
  @AndroidFindBy(id="inputPassword")
  @iOSFindBy(id="inputPassword")
  public MobileElement password_txt;
  //姓名
  @AndroidFindBy(id="inputName")
  @iOSFindBy(id="inputName")
  public MobileElement name_txt;
```

```java
//编程语言控件
@AndroidFindBy(id="input_preferedProgrammingLanguage")
@iOSFindBy(id="input_preferedProgrammingLanguage")
public MobileElement lanquage_sel;
//是否确认注册
@AndroidFindBy(id="input_adds")
@iOSFindBy(id="input_adds")
public MobileElement accept_check;
//注册按钮
@AndroidFindBy(id="btnRegisterUser")
@iOSFindBy(id="btnRegisterUser")
public MobileElement register_btn;

public RegisterVerifyPage register_sucess(String username,String email,String password,
String name, String language,String accept) {
    //页面操作,注册成功后,返回注册验证页面
    log.info("在Register页面, 输入username");
    this.username_txt.sendKeys(username);
    log.info("在Register页面, 输入email");
    this.email_txt.sendKeys(email);
    log.info("在Register页面, 输入password");
    this.password_txt.sendKeys(password);
    log.info("在Register页面, 清空name编辑框");
    this.name_txt.clear();
    log.info("在Register页面, 输入name");
    this.name_txt.sendKeys(name);
    log.info("在Register页面, 单击选择语言,弹出语言选择框");
    this.languge_sel.click();
    log.info("在Register页面, 弹出的语言选择框, 选择语言");
    checkedTextView(language);
    log.info("在Register页面, 选择accept_check");
    if(accept.equals("Yes")){
        if(!this.accept_check.isSelected())
            this.accept_check.click();}
    log.info("在Register页面, 单击注册");
```

```java
    this.register_btn.click();
    return new RegisterVerifyPage(driver);
}


public void checkedTextView(String language){
    //使用class属性选择所有的单选按钮,并存放在一个list中
    @SuppressWarnings("unchecked")
    List<MobileElement> checkedTextViews = (List<MobileElement>) driver.findElementsByCl_
    assName("android.widget.CheckedTextView");
    //使用for循环将list中的每个单选按钮进行遍历,查到name值为"Ruby"的单选按钮,如果该单选按钮未处
    //于选中状态,则调用click方法进行选择
    for(MobileElement checkedTextView:checkedTextViews){
        if(checkedTextView.getAttribute("name").equals(language)){
            if(!checkedTextView.isSelected()){
                checkedTextView.click();}
        }
    }

}
}
```

修改 VerifyRegisterPage 类,如代码清单 8-24 所示。

代码清单 8-24　　　　　　　　　　VerifyRegisterPage 类

```java
package com.shijie.pages;

import org.apache.log4j.Logger;

public class RegisterVerifyPage {
    AndroidDriver<?> driver;
    Logger log = Logger.getLogger(RegisterVerifyPage.class.getName());
    public RegisterVerifyPage(AndroidDriver<?> driver) {
        this.driver = driver;
        PageFactory.initElements(new AppiumFieldDecorator(driver), this);
    }
```

```java
//页面对象识别
@AndroidFindBy(id="label_name_data")
@iOSFindBy(id="label_name_data")
public MobileElement label_name_data;

@AndroidFindBy(id="label_username_data")
@iOSFindBy(id="label_username_data")
public MobileElement label_username_data;

@AndroidFindBy(id="label_password_data")
@iOSFindBy(id="label_password_data")
public MobileElement label_password_data;

@AndroidFindBy(id="label_email_data")
@iOSFindBy(id="label_email_data")
public MobileElement label_email_data;

@AndroidFindBy(id="label_preferedProgrammingLanguage_data")
@iOSFindBy(id="label_preferedProgrammingLanguage_data")
public MobileElement label_preferedProgrammingLanguage_data;

@AndroidFindBy(id="label_acceptAdds_data")
@iOSFindBy(id="label_acceptAdds_data")
public MobileElement label_acceptAdds_data;

//增加验证项
public String get_name_value(){
    log.info("在VerifyRegisterPage,通过输入获取name值");
    return this.label_name_data.getText().toString();
}

public String get_username_value(){
    log.info("在VerifyRegisterPage,通过输入获取username值");
    return this.label_username_data.getText().toString();
}
```

```java
public String get_password_value(){
    log.info("在VerifyRegisterPage,通过输入获取password值");
    return this.label_password_data.getText().toString();
}

public String get_email_value(){
    log.info("在VerifyRegisterPage,通过输入获取email值");
    return this.label_email_data.getText().toString();
}

public String get_preferedProgrammingLanguage_value(){
    log.info("在VerifyRegisterPage,通过输入获取preferedProgrammingLanguage值");
    return this.label_preferedProgrammingLanguage_data.getText().toString();
}

public String get_acceptAdds_value(){
    log.info("在VerifyRegisterPage,通过输入获取acceptAdds值");
    return this.label_acceptAdds_data.getText().toString();
}
}
```

修改 BaseActivity 类,如代码清单 8-25 所示。

代码清单 8-25　　　　　　　　　　BaseActivity 类

```java
package com.shijie.base;
import org.apache.log4j.Logger;
import org.apache.log4j.PropertyConfigurator;
import org.testng.ITestContext;
import org.testng.annotations.AfterClass;
import org.testng.annotations.BeforeClass;
import org.testng.annotations.DataProvider;
import com.shijie.util.Constants;
import com.shijie.util.DataProviderFromExcel;

public class BaseActivity {
```

```java
public static Logger logger = Logger.getLogger(baseActivity.class.getName());

@BeforeClass
/**启动浏览器并打开测试页面*/
public void startTest(ITestContext context) {
    PropertyConfigurator.configure("config/log4j.properties");
    logger.info("---------------测试用例执行开始------------------------------------");
    DesiredCapabilities capabilities = new DesiredCapabilities();
    capabilities.setCapability("deviceName", "Redmi 3");
    capabilities.setCapability("udid", "c1aeae297d72");
    capabilities.setCapability("platformVersion", "5.1.1");
    capabilities.setCapability("platformName", "Android");
    capabilities.setCapability("appPackage", "io.selendroid.testapp");
    capabilities.setCapability("appActivity", ".HomeScreenActivity");
    capabilities.setCapability("noSign", "True");
}

@AfterClass
/**结束测试，关闭浏览器*/
public void endTest() {
    logger.info("---------------测试用例执行结束------------------------------------");
    driver.quit();
}


public AndroidDriver<AndroidElement> getDriver() {
    return driver;
}
}
```

下面查看测试结果。

（1）控制台日志，如图 8-19 所示。

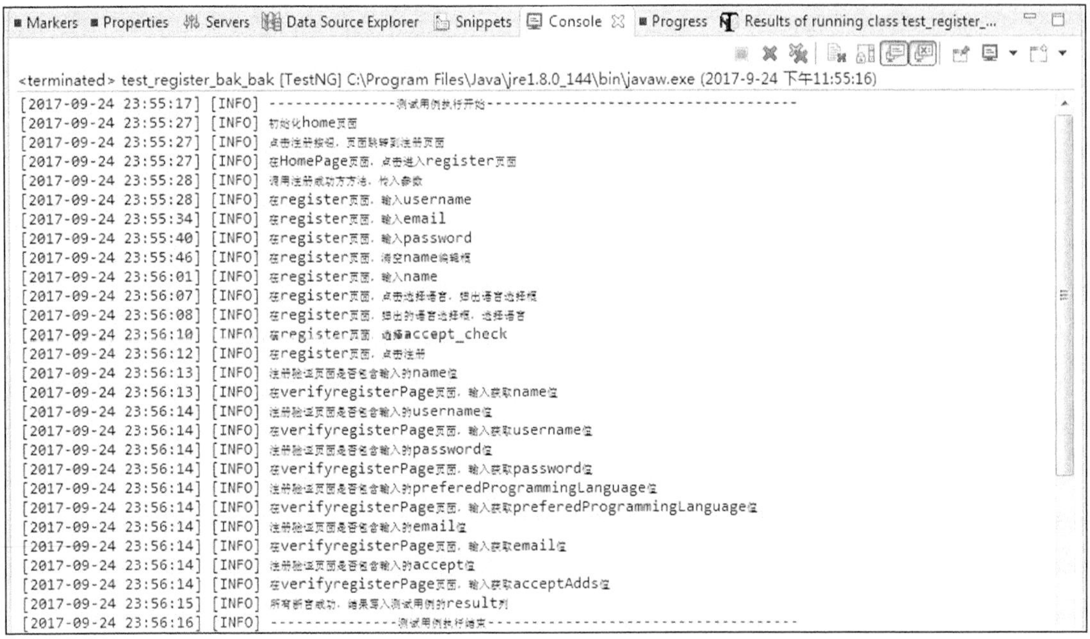

图 8-19 控制台日志

（2）Log 文件日志，如图 8-20 所示。

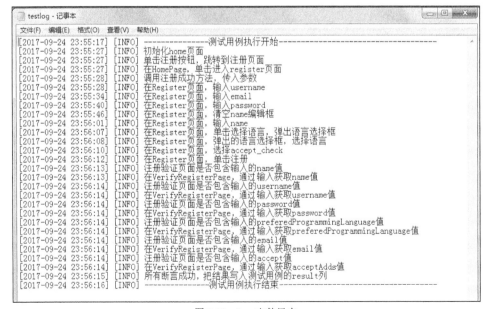

图 8-20 Log 文件日志

8.11　ReportNG 测试报告

测试报告是自动化框架最终展示的书面内容，测试报告的好坏在一定程度上体现了这个自动化框架的好坏。TestNG 本身带有测试报告，不过格式不是很美观。这里采用 ReportNG 来实现测试报告，ReportNG 提供了简单的方式来查看测试结果，并能对结果进行着色，还可以通过修改模板定制内容，修改 CSS 来替换默认的输出样式等。

8.11.1　通过 Maven 导入 ReportNG

在根目录下，新建 TestNG 的运行配置文件 runAll.xml，程序结构如下。

```
AppTest/
|----src/main/java
     |----com.shijie.base/
          |----BaseActivity.java
     |----com.shijie.pages/
          |----HomePage.java
          |----RegisterPage.java
          |----RegisterVerifyPage.java
     |----com.shijie.util/
          |----Contants.java
          |----DataProviderFromCsv.java
|----src/test/java
     |----com.shijie.testScripts/
          |----TestRegister.java
|----data
     |----register.csv
|----config
     |----log4j.properties
|----pom.xml
|----runAll.xml
```

在 Maven 中配置 ReportNG 依赖。添加 ReportNG 的监听器，修改最后的 TestNG 报告，如代码清单 8-26 所示。

代码清单 8-26　　　　　　　　　ReportNG 依赖

```xml
<dependency>
        <groupId>org.uncommons</groupId>
        <artifactId>reportng</artifactId>
        <version>1.1.4</version>
        <scope>test</scope>
        <exclusions>
            <exclusion>
                <groupId>org.testng</groupId>
                <artifactId>testng</artifactId>
            </exclusion>
        </exclusions>
</dependency>
<dependency>
        <groupId>com.google.inject</groupId>
        <artifactId>guice</artifactId>
        <version>3.0</version>
        <scope>test</scope>
</dependency>
```

当配置上述依赖项时，一定要加入 guice 包，否则检查的时候编译不通过，会抛出 ClassNotFoundException 异常。

8.11.2　配置 ReportNG 的监听器

在 runAll.xml 中增加 ReportNG 的监听器，如代码清单 8-27 所示。

代码清单 8-27　　　　　　　　　ReportNG 的监听器

```xml
useDefaultListeners="false"
<listeners>
//加入 ReportNG 的监听器
<listener class-name="org.uncommons.reportng.HTMLReporter" />
<listener class-name="org.uncommons.reportng.JUnitXMLReporter" />
</listeners>
```

其中，要注意几点。配置 userdefaultlisteners 为 false 来禁止 TestNG 产生报告。org.uncommons.reportng.HTMLReporter 用来生成 HTML 格式的报告。org.uncommons.reportng.JUnitXMLReporter 用来生成 JUnit 格式的测试报告。

在 pom.xml 中设置 Report 产生的目录，如代码清单 8-28 所示。

代码清单 8-28　　　　　　　　　　　Report 的设置

```xml
<plugin>
    <groupId>org.apache.maven.plugins</groupId>
    <artifactId>maven-surefire-plugin</artifactId>
    <version>2.19.1</version>
    <configuration>
        <argLine>-Dfile.encoding=UTF-8</argLine>
        <!-- 解决 Maven 内存溢出问题-->
        <argLine>-Xms1024m -Xmx1024m -XX:PermSize=128m -XX:MaxPermSize= 128m</argLine>
        <forkMode>never</forkMode>
        <suiteXmlFiles>
            <suiteXmlFile>testng.xml</suiteXmlFile>
        </suiteXmlFiles>
<!--测试报告路径-->
        <reportsDirectory>result/test-report</reportsDirectory>
    </configuration>
</plugin>
```

8.11.3　执行测试

在 TestNG 的配置文件 runAll.xml 中，输入代码清单 8-29 所示代码。

代码清单 8-29　　　　　　　　　　　TestNG 配置

```xml
<?xml version="1.0" encoding="UTF-8"?>
<suite name="Suite" > <!--parallel="true" 并行-->
    <test name="Test" >
        <classes>
            <class name="com.shijie.testScripts.TestRegister" />
        </classes>
    </test><!--Test-->
</suite><!--Suite-->
```

在 Maven 中指明运行的 TestNG 的配置文件为 runAll.xml，修改 pom.xml 文件，如代码清单 8-30 所示。

代码清单 8-30　　　　　　　　　pom.xml 文件的配置

```xml
<plugin>
    <groupId>org.apache.maven.plugins</groupId>
    <artifactId>maven-surefire-plugin</artifactId>
    <version>2.19.1</version>
        <configuration>
        <!-- 解决用 Maven 执行 test 时日志乱码的问题-->
        <argLine>-Dfile.encoding=UTF-8</argLine>
        <!-- 解决 Maven 内存溢出问题--->
        <argLine>-Xms1024m -Xmx1024m -XX:PermSize=128m -XX:MaxPermSize= 128m</argLine>
        <forkMode>never</forkMode>
        <suiteXmlFiles>
            <suiteXmlFile>runAll.xml</suiteXmlFile>
        </suiteXmlFiles>
    </configuration>
</plugin>
```

在 pom.xml 这个文件上右击，选择 Run As→Maven test，执行测试用例。查看在 "./result/test-report"中产生的测试报告，如图 8-21 所示。

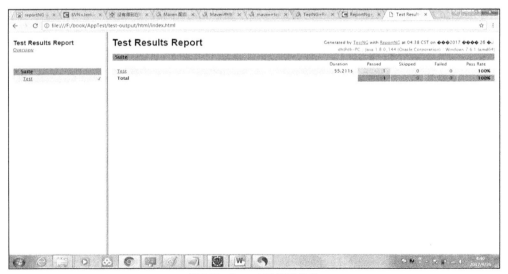

图 8-21　ReportNG 测试报告

8.12 Appium 自启动

上面的自动化测试框架存在一个缺点,就是每次运行时,都要手工启动 Appium 服务。可以使用 AppiumDriverLocalService 在代码中来完成 Appium 服务的启动和关闭,前提是需要安装 Node.js 和 Appium。安装成功后,修改 BaseActivity 类,如代码清单 8-31 所示。

代码清单 8-31　　　　　　　　　BaseActivity 类

```java
package com.shijie.base;
import io.appium.java_client.android.AndroidDriver;
import io.appium.java_client.android.AndroidElement;
import io.appium.java_client.service.local.AppiumDriverLocalService;
import io.appium.java_client.service.local.AppiumServerHasNotBeenStartedLocallyException;
import java.net.MalformedURLException;
import org.apache.log4j.Logger;
import org.apache.log4j.PropertyConfigurator;
import org.testng.ITestContext;
import org.testng.annotations.AfterClass;
import org.testng.annotations.BeforeClass;
import org.testng.annotations.DataProvider;
import com.shijie.util.Constants;
import com.shijie.util.DataProviderFromExcel;
public class BaseActivity {
    public static Logger logger = Logger.getLogger(baseActivity.class.getName());
    AndroidDriver<AndroidElement> driver;
    private static AppiumDriverLocalService service;

    @BeforeClass
    /**启动浏览器并打开测试页面*/
    public void startTest(ITestContext context) throws MalformedURLException {
        logger.info("--------------启动 Appium 服务------------------------------------");
        PropertyConfigurator.configure("config/log4j.properties");
        service = AppiumDriverLocalService.buildDefaultService();
        service.start();
```

```java
      if (service == null || !service.isRunning()) {
        throw new AppiumServerHasNotBeenStartedLocallyException(
          "An appium server node is not started!");
      }

    DesiredCapabilities capabilities = new DesiredCapabilities();
      capabilities.setCapability(MobileCapabilityType.DEVICE_NAME, "Redmi 3");
      capabilities.setCapability(MobileCapabilityType.UDID, "c1aeae297d72");
      capabilities.setCapability(MobileCapabilityType.PLATFORM_VERSION, "5.1.1");
      capabilities.setCapability(MobileCapabilityType.PLATFORM_NAME, "Android");
      capabilities.setCapability(AndroidMobileCapabilityType.APP_PACKAGE,"io.selendroid.testapp");
      capabilities.setCapability(AndroidMobileCapabilityType.APP_ACTIVITY,".HomeScreenActivity");
      capabilities.setCapability(AndroidMobileCapabilityType.UNICODE_KEYBOARD, "True");
      capabilities.setCapability(AndroidMobileCapabilityType.NO_SIGN, "True");
      driver = new AndroidDriver<>(new URL("http://127.0.0.1:4723/wd/hub"), capabilities);
    }


    @AfterClass
    /**结束测试并关闭浏览器*/
    public void endTest() {
      logger.info("---------------测试用例执行结束------------------------------------");
    driver.quit();
      if (service != null) {
        service.stop();
      }
    }


    public AndroidDriver<AndroidElement> getDriver() {
      // TODO Auto-generated method stub
      return driver;
    }
}
```

若使用上述代码，在测试执行前，就无须手工启动 Appium GUI 端。

本例中使用 CSV 作为数据驱动，搭建起一个基于 Appium 的完整测试框架，读者可以根据自己掌握的知识并结合项目的实际情况，编写符合要求的自动化测试框架。一般 App 测试离不开 Web 测试，将 App 自动化测试和 Web 自动化测试结合起来，才能真正形成一个闭环，完善功能测试。读者可以借鉴此框架，在其上增加 Web 自动化测试，增加 Android 性能测试数据的获取，增加 iOS 的自动化测试等功能，这样的自动化框架功能将更加强大。

第 9 章

Appium 关键字驱动测试框架封装实战

前面第 8 章的自动化框架主要是基于数据驱动的，接下来本章主要讲解关键字驱动框架，并用该框架继续完善和扩充上一章的相关功能。

关键字驱动测试也称为"表格驱动测试"或"操作名测试"。关键字驱动框架的基本工作是将测试用例分成 4 个不同的部分：一是测试步骤（Test Step），二是测试步骤中的对象（Object），三是测试对象执行的动作（Action），四是测试对象需要的数据（Test Data）。

本章将测试逻辑按照这个关键字进行分解，形成数据文件，存放在 Excel 文件中，同时创建测试用例调度文件来管理这些测试逻辑。

9.1 搭建测试框架

搭建测试框架的步骤如下。

（1）新建 Maven 项目，命名为 KeyWordAppTest，并按照第 8 章的描述，基本类似于数据驱动框架，在项目中导入 Maven、TestNG、Log4j 和 ReportNG 的依赖包。

（2）在项目中新建 3 个包（com.shijie.base、com.shijie.util 和 com.shijie.testScripts），以及 4 个文件夹（apps.config、data 和 result），程序结构如下。

```
KeyWordAppTest/
|----src/main/java/
     |----com.shijie.base/
     |----com.shijie.util/
|----src/test/java/
     |----com.shijie.testScripts/
|----data/
|----result/
|----config/
|----pom.xml
```

3 个包的功能如下。

- com.shijie.base，用于实现可复用的复杂的业务逻辑封装。
- com.shijie.testScripts，具体的测试类。
- com.shijie.util，公共类库，实现配置文件的读取、Excel 数据驱动、页面操作重新封装等。

文件夹的功能如下。

- apps，存放待测 App。
- config，存放配置文件，包括驱动配置文件、日志配置文件以及常用路径配置文件。
- data，测试数据，存放 Excel 测试用例文件。
- result，结果文件，包括 log 目录、screenshot 目录以及 test-report 目录。

（3）被测功能和最开始的测试脚本参考 8.6.2 节。在 com.shijie.testCase 下面新建测试用例 TestRegister 类。脚本内容参考代码清单 8-3。

9.2 代码优化

在 com.shijie.base 包中创建 BaseActivity 类，用于将 TestRegister 测试类在测试前后的环境准备和环境恢复单独提取出来，代码结构如下。

```
KeyWordAppTest/
|----src/main/java/
     |----com.shijie.base/
          |----BaseActivity.java
     |----com.shijie.util/
          |----Contants.java
|----src/test/java/
     |----com.shijie.testScripts/
          |----TestRegister.java
|----data/
|----result/
|----config/
|----pom.xml
```

其次是针对 Desired Capabilities 的设置，由于每个移动设备都需要配置，因此为了方便测试人员配置不同的移动设备并将配置提取出来，在 com.shijie.util 包中创建 Constants 类，用于存放配置文件。以增加 Constants 类中关于红米手机 Desired Capabilities 的配置为例来配置 Constants 类，如代码清单 9-1 所示。

代码清单 9-1　　　　　　　　　　Constants 类

```java
public class 测试手机_红米{
    public static final String deviceName = "Redmi 3";
    public static final String udid = "c1aeae297d72";
    public static final String platformVersion = "5.1.1";
    public static final String platformName = "Android";
    public static final String appPackage = "io.selendroid.testapp";
    public static final String appActivity = ".HomeScreenActivity";
    public static final String unicodeKeyboard = "True";
    public static final String noSign = "True";
}
```

BaseActivity 类的具体代码如代码清单 9-2 所示。

代码清单 9-2　　　　　　　　　　BaseActivity 类

```java
package com.shijie.base;
import io.appium.java_client.android.AndroidDriver;
import io.appium.java_client.android.AndroidElement;
import java.net.URL;
import org.apache.log4j.Logger;
import org.apache.log4j.PropertyConfigurator;
import org.openqa.selenium.remote.DesiredCapabilities;
import org.testng.annotations.AfterClass;
import org.testng.annotations.BeforeClass;
import org.testng.annotations.Parameters;
import com.shijie.util.Constants;


public class BaseActivity {
    public static Logger logger = Logger.getLogger(baseActivity.class.getName());
    public static AndroidDriver<AndroidElement> driver;
    @BeforeClass
    public void SetUp() throws Exception  {
        PropertyConfigurator.configure("config/log4j.properties");
        logger.info("---------------测试用例执行开始--------------------------------");
        // TODO Auto-generated method stub
```

```java
        DesiredCapabilities capabilities = new DesiredCapabilities();
        capabilities.setCapability("deviceName", Constants.测试手机_红米.deviceName);
        capabilities.setCapability("udid", Constants.测试手机_红米.udid);
        capabilities.setCapability("platformVersion", Constants.测试手机_红米.platformVersion);
        capabilities.setCapability("platformName", Constants.测试手机_红米.platformName);
        capabilities.setCapability("appPackage", Constants.测试手机_红米.appPackage);
        capabilities.setCapability("appActivity", Constants.测试手机_红米.appActivity);
        capabilities.setCapability("unicodeKeyboard", Constants.测试手机_红米.unicodeKeyboard);
        capabilities.setCapability("noSign", Constants.测试手机_红米.noSign);
        driver = new AndroidDriver<>(new URL("http://127.0.0.1:4723/wd/hub"), capabilities);
    }


    @AfterClass
    /**结束测试并关闭*/
    public void endTest() {
        logger.info("----------------测试用例执行结束----------------------------------");
        driver.quit();
    }


    public AndroidDriver<AndroidElement> getDriver() {
        return driver;
    }
}
```

9.3 关键字驱动

CSV 和 Excel 都是自动化测试中常用的数据展示方式。在上一节中，数据源的获取是通过 CSV 文件完成的，这一节将采用 Excel 文件格式作为数据源，在项目中通过解析 Excel 表，来获取测试数据。

（1）在 pom.xml 中配置 Excel 依赖包，配置完成后的 pom.xml 如代码清单 9-3 所示。

代码清单 9-3　　　　　　　在 Maven 中配置 Excel 依赖包

```xml
<dependency>
    <groupId>org.apache.poi</groupId>
```

```xml
        <artifactId>poi</artifactId>
        <version>3.17</version>
    </dependency>

    <dependency>
        <groupId>org.apache.poi</groupId>
        <artifactId>poi-ooxml</artifactId>
        <version>3.17</version>
    </dependency>
```

配置完成后，Maven 会自动下载 poi 包到本地。这里值得注意的是，poi 有两个不同的 Jar 包——poi 和 poi-ooxml，二者分别处理 Excel 2003 及以下版本和 Excel 2007 及以上版本。本章只针对 Excel 2007 及以上版本，poi 依赖包的配置仅供参考。

（2）在测试项目根目录下创建 data 文件夹，该文件夹用来放置测试数据，创建测试文件 register.xlsx。

（3）测试用例数据。

①通常，一个业务流程会包含多个测试用例，创建测试任务调度 Sheet 页，如图 9-1 所示。

A	B	C	D
testCaseName	testCaseDetail	isRun	result
register	完成完整的注册测试用例	√	

图 9-1　测试任务调度 Sheet 页

第 1 列为 testCaseName，即测试用例名。测试用例名字段直接对应测试用例所在的 Sheet 页的名字。

第 2 列为 testCaseDetail，即测试用例的详细描述。

第 3 列为 isRun，表示测试用例是否运行，"√"表示运行，"×"或者其他符号表示不运行。如图 9-1 所示，测试用例"register"对应的"isRun"字段为"√"，表示会运行 Sheet 页中名称为"register"的测试用例。

第 4 列为 result，表示测试用例执行结果。整个测试用例中所有测试步骤执行完成后，如果所有测试步骤都运行成功，result 列结果为"true"，如果有一个步骤执行失败，结果为"false"。

②对于测试用例的具体内容所在的 Sheet 页，即具体的测试步骤，内容由测试工程师输

入，具体内容见表9-1。

表9-1 测试用例文件数据

testCaseID	testStepID	testStepDetail	objectName	inspector	actionStep	data	result
register1	testStep1	单击注册文件夹	首页-注册	io.selendroid.testapp:id/startUserRegistration	click	null	—
register1	testStep2	等待页面跳转	首页-注册		waitForLoadingActivity	.RegisterUserActivity	—
register1	testStep3	输入用户名	注册页-用户名	io.selendroid.testapp:id/inputUsername	input	sh***	—
register1	testStep4	输入邮箱	注册页-邮箱	io.selendroid.testapp:id/inputEmail	input	sh***@126.com	—
register1	testStep5	输入密码	注册页-密码	io.selendroid.testapp:id/inputPassword	input	123456	—
register1	testStep6	输入姓名	注册页-姓名	io.selendroid.testapp:id/inputName	input	daming	—
register1	testStep7	单击选择语言	注册页-选择语言	io.selendroid.testapp:id/input_preferedProgrammingLanguage	click	null	—
register1	testStep8	选择语言	注册页-语言	PHP	click	null	—
register1	testStep9	选择是否接受	注册页-是否接受	io.selendroid.testapp:id/input_adds	click_radio	yes	—
register1	testStep10	单击注册	注册页-单击注册	io.selendroid.testapp:id/btnRegisterUser	click	null	—
register1	testStep11	等待页面跳转	注册页-单击注册		waitForLoadingActivity	.VerifyUserActivity	—
register1	testStep12	验证用户名	验证页-用户名	io.selendroid.testapp:id/label_username_data	verify	sh***	—
register1	testStep13	验证邮箱	验证页-邮箱	io.selendroid.testapp:id/label_email_data	verify	sh***@126.com	—
register1	testStep14	验证密码	验证页-密码	io.selendroid.testapp:id/label_password_data	verify	123456	—
register1	testStep15	验证姓名	验证页-姓名	io.selendroid.testapp:id/label_name_data	verify	daming	—
register1	testStep16	验证语言	验证页-语言	io.selendroid.testapp:id/label_preferedProgrammingLanguage_data	verify	PHP	—
register1	testStep17	验证选择是否接受	验证页-是否接受	io.selendroid.testapp:id/label_acceptAdds_data	verify	yes	—

第1列为testCaseID，即测试用例编号。

第2列为testStepID，即测试步骤编号。

第3列为testStepDetail，即测试步骤详细描述。

第4列为objectName，即对象名称。

第5列为inspector，即对象识别方式。

第6列为actionStep，即操作步骤。

第7列为data，即操作数据。

第8列为result，即执行结果。

每读取一行测试步骤，就执行对应的方法。执行结束后，将执行结果写入对应的 result 列中。

（4）在 com.shijie.util 包中创建 DataProviderFromExcel 类，用于实现对 Excel 文件的操作（包括文件的路径和具体的 Sheet 页，读取指定单元格的值，写入内容到指定单元格等）。代码结构如下。

```
KeyWordAppTest/
|----src/main/java/
     |----com.shijie.base/
          |----BaseActivity.java
     |----com.shijie.util/
          |----Contants.java
          |----DataProviderFromExcel.java
|----src/test/java/
     |----com.shijie.testScripts/
|----data/
|----result/
|----config/
|----pom.xml
```

DataProviderFromExcel 类的具体内容如代码清单 9-4 所示。

代码清单 9-4　　　　　　　　　DataProviderFromExcel 类

```java
package com.shijie.util;

import java.io.File;

import java.io.FileInputStream;

import java.io.FileOutputStream;

import org.apache.poi.ss.usermodel.CellType;

import org.apache.poi.xssf.usermodel.XSSFCell;

import org.apache.poi.xssf.usermodel.XSSFRow;

import org.apache.poi.xssf.usermodel.XSSFSheet;

import org.apache.poi.xssf.usermodel.XSSFWorkbook;

public class DataProviderFromExcel {

    private static File file = null;
```

```java
private static XSSFWorkbook work = null;
private static XSSFSheet sheet = null;
private static XSSFRow row = null;
private static XSSFCell cell = null;

/**
 * 初始化 Excel 文档,设定待操作的文件路径和 Sheet 页
 */
public static void getExcel(String filePath) throws Exception{
//String fileAbsolutePath = getFileAbsolutePath(filePath, fileName);
    file = new File(filePath);
    FileInputStream inputStream = new FileInputStream(file);
    work = new XSSFWorkbook(inputStream);
      if(null == work){
          throw new Exception("创建 Excel 工作簿为空!");
      }
}


/**
 * 读取指定单元格的值
 */
public static String getCellData(String sheetName, int rowNum, int colNum) throws Exception{
    sheet = work.getSheet(sheetName);
    try{
        cell = sheet.getRow(rowNum).getCell(colNum);
        String cellData = getCellValue(cell);
            return cellData;
        }catch(Exception e){
            return "";
        }
}

/**
 * 根据 Excel 中格式的不同来读取不同格式的值
```

```java
 */
private static String getCellValue(XSSFCell cell) {
    String strCell = "";
    if(cell.getCellTypeEnum() == CellType.STRING){
        strCell = cell.getStringCellValue();
    }
    else if(cell.getCellTypeEnum() == CellType.NUMERIC){
        strCell = String.valueOf(cell.getNumericCellValue());
        strCell = strCell.split(".")[0];
    }
    else if(cell.getCellTypeEnum() == CellType.BOOLEAN){
        strCell = String.valueOf(cell.getBooleanCellValue());
    }
    else if(cell.getCellTypeEnum() == CellType.BLANK){
        strCell = "";
    }
    else {
        strCell = "";
    }
        return strCell;
}

/**
 * 向指定的单元格写入数据
 */
public static void setCellData(int rowNum, int colNum, boolean result, String sheetName, String filePath) throws Exception{
    sheet = work.getSheet(sheetName);
    try{
        row = sheet.getRow(rowNum);
        cell = row.getCell(colNum);
        if(cell == null){
            cell = row.createCell(colNum);
            cell.setCellValue(result);
        }
```

```
            else{
                cell.setCellValue(result);
            }
        FileOutputStream outputStream = new FileOutputStream(filePath);
        work.write(outputStream);
        outputStream.close();
    }catch(Exception e){
        throw e;
    }
    }
/**
*获取指定 Sheet 页的单元格行数
*/
    public static int getAllRowNum(String sheetName) {
        // TODO Auto-generated method stub
        sheet = work.getSheet(sheetName);
        return sheet.getLastRowNum();
    }
}
```

（5）修改 Constants 类，如代码清单 9-5 所示，在"测试 Excel 文件"类中写入作为数据来源的 Excel 的路径。

代码清单 9-5　　　　　　　　　　Constants 类

```
public class 测试Excel文件{
    public static final String filepath = "D:\\Users\\dh\\workspace\\keywordAppTest\\data\\register.xlsx";
}
```

（6）测试调度文件、测试用例文件所在的 Sheet 页中，每列都有具体的含义，Excel 是靠行值和列值来确定具体单元格的。为了方便调用，在 Constants 类中完成每列的常量设定，如代码清单 9-6 所示。例如，测试调度文件 Sheet 中使用的 suite_testCaseName 对应第 0 列，在测试用例文件所在的 Sheet 中使用的 Col_testcaseID 对应第 0 列。

代码清单 9-6　　　　　　　　　　Constants 类

```
package com.shijie.util;

public class Constants {
```

```java
public class 测试Excel文件{
    public static final String filepath = "D:\\Users\\dh\\workspace\\keywordAppTest\\data\\register.xlsx";
}

public class 测试调度文件{
    public static final String suite_sheet = "Main";
    public static final int suite_testCaseName = 0;
    public static final int suite_testCaseDetail = 1;
    public static final int suite_isRun = 2;
    public static final int suite_result = 3;
}

public class 测试用例文件{
    public static final int Col_testcaseID = 0;
    public static final int Col_testStepID = 1;
    public static final int Col_inspector = 4;
    public static final int Col_data = 6;
    public static final int Col_result = 7;
    public static final int Col_actionStep = 5;
}
```

9.4 页面元素的封装

在 Appium 数据驱动测试框架中对于定位元素对象采用了 PageFactory 设计模式。优点是跨平台代码简洁，代码的编写和维护效率更高。缺点也是显而易见的，如在前面数据驱动的自动化测试框架中，有个很大的不足之处就是等待的时间是固定的。PageFactory 基于注解的方式对控件进行识别，无法根据需要自定义封装。

本节介绍自定义封装页面元素识别对象类 findElement。这个对象可以完成等待操作和对象识别。

在 Web 自动化测试中使用 WebDriverWait 的 until 和 until not 来完成等待操作，这里封装 findElement 对象来完成同样的功能。该方法需要的参数是 driver 和对象识别的方法，具体的实现功能为每隔一段时间检查控件是否被识别。把识别次数和间隔时间写入 Constants 类中，如代码清单 9-7 所示。修改 Constants 类，增加识别次数和间隔时间。如果在整个识别次数内未找到对象，则整个测试用例置为 False。本例中，控件的识别次数为 5 次，间隔时间为 1000ms。

代码清单 9-7　　　　　　　　　　Constants 类

```
public class 测试控件{
    public static final int elementInspectCount = 5;
    public static final int elementInspectInterval = 1000;
    public static final int activityInspectCount =5;
    public static final int activityInspectInterval =5;
}
```

在测试执行的过程中发现，对于元素的识别，通过 ID、Name 或者 XPath 进行识别，都是对字符串进行传入。通过解析字符串调用不同的识别方式，如果通过 ID 进行识别，字符串中会包含 ":/id" 字样；如果通过 Xpath 进行识别，字符串会以 "//" 进行开头；如果通过 Name 进行识别，会调用 "findElementByAndroidUIAutomator"。下面在 com.shijie.util 包中创建 FindElement 类以完成对象的识别，返回值为 AndroidElement 对象。

程序结构如下。

```
KeyWordAppTest/
|----src/main/java/
    |----com.shijie.base/
        |----BaseActivity.java
    |----com.shijie.util/
        |----Contants.java
        |----DataProviderFromExcel.java
        |----FindElement.java
|----src/test/java/
    |----com.shijie.testScripts/
|----data/
|----result/
|----config/
|----pom.xml
```

FindElement 类的具体内容如代码清单 9-8 所示。

代码清单 9-8　　　　　　　　　　　　FindElement 类

```java
package com.shijie.util;
import org.apache.log4j.Logger;
import com.shijie.testScripts.testRegister;
import io.appium.java_client.MobileElement;
import io.appium.java_client.android.AndroidDriver;
import io.appium.java_client.android.AndroidElement;
public class FindElement{
    static Logger log = Logger.getLogger(FindElement.class.getName());


/*对象是被函数进行二次封装
*/
    public static AndroidElement findElementbyType (AndroidDriver<?> driver, String
    controlInfo)throws Exception{
        MobileElement element = null;
        if(controlInfo.startsWith("//")){
            element = (MobileElement) driver.findElementByXPath(controlInfo);
        }
        else if(controlInfo.contains(":id/")){
            element = (MobileElement) driver.findElementById(controlInfo);
        }else{
            try{
                element = (MobileElement) driver.findElementByAndroidUIAutomator("text
                (\""+controlInfo+"\")");
            }catch(Exception e){
                element = (MobileElement) driver.findElementByClassName(controlInfo);
            }
        }
        return (AndroidElement) element;
    }

    public static MobileElement findElement(AndroidDriver<?> driver ,String controlInfo)
    throws Exception{
        int elementInspectCount,elementInspectInterval;
```

```java
        elementInspectCount = Constants.测试控件.elementInspectCount;
        elementInspectInterval = Constants.测试控件.elementInspectInteval;
        MobileElement element = null;
        for(int i =0; i<elementInspectCount;i++){
            Thread.sleep(elementInspectInterval);
            try{
                element = findElementbyType(driver, controlInfo);
                log.info("已经找到元素对象");
                return element;
            }catch(Exception e){
                log.info("控件未出现! Waitting.........1s");
                continue;
            }
        }
        log.info("多次查找未找到元素控件,这时测试结果置位false");
        testRegister.testResult = false;
        throw new IllegalArgumentException("在指定时间内未找到测试对象");
    }
}
```

测试脚本中大部分控件的查找都只需要调用该方法即可实现。此外，对于查找一组控件，也可以采用同样的方法。

9.5 测试操作的封装

对于 FindElement 类识别的控件元素 AndroidElement 对象，需要调用具体的操作方法完成一系列独立的测试步骤，如单击、输入等。在 com.shijie.util 中创建 Action 类，程序结构如下。

```
KeyWordAppTest/
|----src/main/java/
     |----com.shijie.base/
          |----BaseActivity.java
     |----com.shijie.util/
          |----Contants.java
```

```
            |----DataProviderFromExcel.java
            |----FindElement.java
            |----Action.java
|----src/test/java/
    |----com.shijie.testScripts/
|----data/
|----result/
|----config/
|----pom.xml
```

为了保持参数的一致性,所有操作方法传入的参数都为 MobileElement mobileElement, String data。有些参数可能在该方法中没有意义,如 click 方法中传入的 data 参数,没有实质上的意义,只是为了保持代码的一致性。

下面介绍具体的操作方法。

1. click 方法

如代码清单 9-9 所示,单击识别控件,如果出现异常,指定间隔时间后再次单击控件。间隔时间可以在 Constants 类的"测试控件"的 elementInspectInterval 中进行设置。

代码清单 9-9　　　　　　　　　　click 方法

```java
/**实现单击操作,这里的data没有任何以意义,只是为了维护脚本的完整性
*/
public void click(MobileElement mobileElement, String data){
    try{
        mobileElement.click();
    }catch(Exception e){
        try {
            Thread.sleep(Constants.测试控件.elementInspectInterval);
        } catch (InterruptedException e1) {
            testRegister.testResult = false;
            e1.printStackTrace();
        }
        mobileElement.click();
    }
}
```

2. click_radio 方法

如代码清单 9-10 所示，radio 控件有选中和非选中两种状态，可以获取所需要的状态，并判断控件目前的状态。当需要选中时，若目前为选中状态则不执行操作，若为未选中状态则执行单击选中操作。

代码清单 9-10　click_radio 方法

```java
/**针对单元框进行单击，单击前判断是否已经处于选择状态
*/
public void click_radio(MobileElement mobileElement, String data){
    try{
//System.out.println("wahaha");
        if(data.toLowerCase().equals("yes")){
            if(!mobileElement.isSelected())
                mobileElement.click();
        }
    }catch(Exception e){
        testRegister.testResult = false;
        e.printStackTrace();
    }
}
```

3. input 方法

如代码清单 9-11 所示，首先清空编辑框的数据，然后进行输入。

代码清单 9-11　input 方法

```java
/**在编辑框输入指定的数据
*/
public void input(MobileElement mobileElement,String data){
    try{
        this.click(mobileElement,data);
        mobileElement.clear();
        mobileElement.sendKeys(data);
    }
    catch(Exception e){
```

```
            testRegister.testResult = false;
            e.printStackTrace();
        }
    }
```

4. verify 方法

如代码清单 9-12 所示，在本例中，输入注册信息后会弹出验证页面，对于输入数据进行验证，如果一致表示输入成功。

代码清单 9-12　　　　　　　　　　　verify 方法

```
/**验证操作
*/
    public void verify(MobileElement mobileElement,String data){
        String actualResult;
        try{
            actualResult = mobileElement.getAttribute("text");
            if(!actualResult.equals(data)){
                testRegister.testResult = false;
            }
        }catch(Exception e){
            testRegister.testResult = false;
        }
    }
```

5. waitForLoadingActivity 方法

当页面跳转的时候，为了查看是否跳转到指定的 Activity，可以通过 adb 命名查看当前的 Activity 值。

```
db shell dumpsys window windows | findstr "mFocusedApp"
```

返回结果为当前的 Activity 值，如图 9-2 所示。

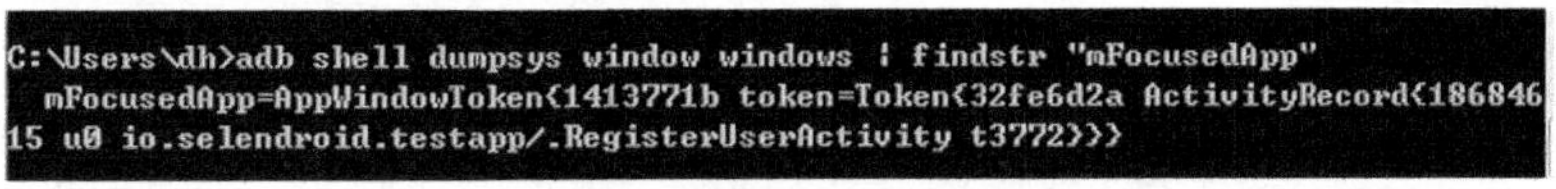

```
C:\Users\dh>adb shell dumpsys window windows | findstr "mFocusedApp"
  mFocusedApp=AppWindowToken{1413771b token=Token{32fe6d2a ActivityRecord{186846
15 u0 io.selendroid.testapp/.RegisterUserActivity t3772}}}
```

图 9-2　返回当前 Activity 值

循环指定的次数，每次间隔一段时间后查看当前的 Activity 值。如果一致，则表示跳转成功；如果不一致，则表示跳转失败。

循环次数可以在 Constants 的类"测试控件"中 elementInspectCount 进行设置，间隔时间在 Constants 类的"测试控件"的 elementInspectInterval 中进行设置，如代码清单 9-13 所示。

代码清单 9-13　　　　　　　　　　Constants 类

```java
public class 测试控件{
    public static final int elementInspectCount = 5;
    public static final int elementInspectInterval = 1000;
    public static final int activityInspectCount =5;
    public static final int activityInspectInterval =1000;
}
```

查找 Activity 的次数为 5 次，时间间隔为 1000ms。waitForLoadingActivity 方法的具体内容如代码清单 9-14 所示。

代码清单 9-14　　　　　　　　waitForLoadingActivity 方法

```java
/**等待Activity跳转
*/
public static void waitForLoadingActivity(MobileElement mobileElement,String data) throws InterruptedException {
        Thread.sleep(3000);
        log.info(driver.currentActivity());
        int activityInspectCount,activityInspectInterval;
        activityInspectCount = Constants.测试控件.activityInspectCount;
        activityInspectInterval = Constants.测试控件.activityInspectInterval;
    int i =0;
    Thread.sleep(activityInspectInterval);
    while (i<activityInspectCount) {
      try {
        if (data.contains(driver.currentActivity())) {
            log.info(data+"出现! ");
            break;
        }
        else
        {
            log.info(data+"未出现! Waiting.........1s");
```

```
                    Thread.sleep(activityInspectInterval);
                    i++;
                }
            } catch (Exception e) {
                i++;
                log.info("尝试"+activityInspectCount+"次,"+data+",未出现! ");
                testRegister.testResult = false;
            }
        }
    }
}
```

完整 Action 类的内容如代码清单 9-15 所示。

代码清单 9-15   Action 类

```java
package com.shijie.util;

import com.shijie.testScripts.testRegister;
import io.appium.java_client.android.AndroidDriver;
import io.appium.java_client.MobileElement;
import io.appium.java_client.android.AndroidElement;
import io.appium.java_client.android.AndroidKeyCode;
import org.apache.log4j.Logger;

public class Action{
    static Logger log = Logger.getLogger(Action.class.getName());
    private static AndroidDriver<AndroidElement> driver;
//public AndroidDriver<AndroidElement> driver;
    public Action(AndroidDriver<AndroidElement> driver) {
        this.driver = driver;
    }
    /**实现单击操作,这里的data没有任何意义,只是为了维护脚本的完整性
    */
    public void click(MobileElement mobileElement, String data){
        try{
            mobileElement.click();
        }catch(Exception e){
            try {
                Thread.sleep(Constants.测试控件.elementInspectInterval);
```

```java
            } catch (InterruptedException e1) {
                testRegister.testResult = false;
                e1.printStackTrace();
            }
            mobileElement.click();
        }
    }


    /**针对单元框进行单击,单击前判断是否已经处于选择状态
    */
    public void click_radio(MobileElement mobileElement, String data){
        try{
//          System.out.println("wahaha");
            if(data.toLowerCase().equals("yes")){
                if(!mobileElement.isSelected())
                    mobileElement.click();
            }
        }catch(Exception e){
            testRegister.testResult = false;
            e.printStackTrace();
        }
    }


    /**对编辑框输入指定的数据
    */
    public void input(MobileElement mobileElement,String data){
        try{
            this.click(mobileElement,data);
            mobileElement.clear();
            mobileElement.sendKeys(data);
        }
        catch(Exception e){
            testRegister.testResult = false;
            e.printStackTrace();
```

            }
        }

/**后退操作
*/
    public void back(MobileElement mobileElement,String data){
        driver.pressKeyCode(AndroidKeyCode.BACK);
    }

/**验证操作
*/
    public void verify(MobileElement mobileElement,String data){
        String actualResult;
        try{
            actualResult = mobileElement.getAttribute("text");
            if(!actualResult.equals(data)){
                testRegister.testResult = false;
            }
        }catch(Exception e){
            testRegister.testResult = false;
        }
    }

/**等待Activity跳转
*/
    public static void waitForLoadingActivity(MobileElement mobileElement,String data)
    throws InterruptedException {
        Thread.sleep(3000);
        log.info(driver.currentActivity());
        int activityInspectCount,activityInspectInterval;
        activityInspectCount = Constants.测试控件.activityInspectCount;
        activityInspectInterval = Constants.测试控件.activityInspectInterval;
        int i =0;
        Thread.sleep(activityInspectInterval);
        while (i<activityInspectCount) {

```
            try {
                if (data.contains(driver.currentActivity())) {
                    log.info(data+"出现！ ");
                    break;
                }
                else
                {
                    log.info(data+"未出现! Waiting.........1s");
                    Thread.sleep(activityInspectInterval);
                    i++;
                }
            } catch (Exception e) {
                i++;
                log.info("尝试"+activityInspectCount+"次,"+data+",未出现！ ");
                testRegister.testResult = false;
            }
        }
    }
}
```

Action 的封装只针对本次测试用例所使用到的操作，对于其他控件的封装可以根据需要进行调整。这里对于 waitForLoadingActivity 只是做了简单的封装，较复杂的场景建议使用 WebDriverWait 进行二次封装。

9.6 执行测试

完成以上操作后，修改 TestRegister，如代码清单 9-16 所示，用于读取测试调度文件和测试用例，调用测试步骤，完成测试任务。具体的执行步骤如下。

（1）打开测试调度文件。

（2）逐行读取测试用例，通过"isRun"字段值判断该条测试用例是否运行。如果为"×"，则继续读取下一条测试用例；如果为"√"，则读取"testCaseName"列的值，找到同名的 Sheet 页，读取测试步骤。

（3）在测试用例 Sheet 页中，逐条读取测试步骤，调用测试方法进行运行。如果测试步骤执行成功，则 result 字段值为"TRUE"；否则，为"FALSE"。

（4）在测试用例 Sheet 页，通过 Inspector 列的值定位页面控件元素，通过 actionStep 指定的操作方法进行操作。如果需要输入，在 data 列输入值。

（5）在测试用例的 Sheet 页，如果所有的测试步骤执行成功，则测试调度文件的 result 列值为"TRUE"；如果有一个测试步骤执行失败，就跳出测试用例的执行，同时将测试调度文件中 result 列的值设为"FALSE"。

代码清单 9-16　　　　　　　　　　TestRegister 类

```java
package com.shijie.testScripts;

import com.shijie.base.baseActivity;
import com.shijie.util.Action;
import com.shijie.util.FindElement;
import com.shijie.util.Constants;
import io.appium.java_client.MobileElement;
import com.shijie.util.DataProviderFromExcel;
import java.lang.reflect.Method;
import org.testng.Assert;
import org.testng.annotations.Test;
import org.apache.log4j.Logger;

public class TestRegister extends BaseActivity {
//public class testRegister {
    Logger log =  Logger.getLogger(testRegister.class.getName());
    public static Method method[];
    //对象识别关键字
    public static String inspector ;
    //数据
    public static String data ;
    //操作
    public static String actionstep ;
    //定义类
    public static  Action action ;
    public static MobileElement mobileElememnt ;
    public static boolean testResult;
```

```java
public static String filePath;

@Test
public void test_register_sucess() throws Exception{
    action= new Action(getDriver());
    method = action.getClass().getMethods();
    //定义excel文件的路径
    String filePath = Constants.测试excel文件.filepath;
    DataProviderFromExcel.getExcel(filePath);
    String fileSheet = Constants.测试调度文件.suite_sheet;
    //获取测试集合中测试用例的总数
    int testSuiteAllNum = DataProviderFromExcel.getAllRowNum(fileSheet);
    //循环测试调度文件
    for(int testSuiteNum =1 ;testSuiteNum<= testSuiteAllNum;testSuiteNum++){
        //获取测试用例名,直接关联待用例所在的sheet名
        String testCaseName = DataProviderFromExcel.getCellData(Constants.测试调度文件.suite_sheet, testSuiteNum, Constants.测试调度文件.suite_testCaseName).trim();
        //判断测试用例是否运行
        String isRun = DataProviderFromExcel.getCellData(Constants.测试调度文件.suite_sheet, testSuiteNum, Constants.测试调度文件.suite_isRun).trim();
        //获取测试用例详细的描述,只用来输出日志
        String testCaseDetail = DataProviderFromExcel.getCellData(Constants.测试调度文件.suite_sheet, testSuiteNum, Constants.测试调度文件.suite_testCaseDetail).trim();
        //如果isRun的值为"√",则执行指定sheet页的测试步骤,sheet名与testCaseName相同
        if(isRun.equals("√")){
            //测试执行结果默认为失败
            log.info("运行测试用例:测试用例名称:"+testCaseName+";测试用例详细描述:"+testCaseDetail);
            testResult = true;
            int testCaseAllNum = DataProviderFromExcel.getAllRowNum(testCaseName);
            log.info("测试步骤数: " + testCaseAllNum);
            for(int testcaseNum =1 ;testcaseNum<= testCaseAllNum;testcaseNum++){
                //获取识别方式
                inspector = DataProviderFromExcel.getCellData(testCaseName, testcaseNum, Constants.测试用例文件.Col_inspector).trim();
                //获取操作数据
                data = DataProviderFromExcel.getCellData(testCaseName, testcaseNum,
```

```
                Constants.测试用例文件.Col_data).trim();
            //获取操作方式
            actionstep = DataProviderFromExcel.getCellData(testCaseName,
            testcaseNum, Constants.测试用例文件.Col_actionStep).trim();
            //识别元素
            mobileElememnt = null;
            if(!inspector.isEmpty()){
                mobileElememnt = FindElement.findElement(driver, inspector);
            }
            log.info("执行测试步骤: 识别方式: "+inspector+"; 操作: "+actionstep+";
            测试数据: "+data);
            execute_Actions(testcaseNum, testCaseName);
            if(testResult==false){
                log.info("测试用例执行结果为false");
                DataProviderFromExcel.setCellData(testSuiteNum, Constants.
                测试调度文件.suite_result, false, fileSheet, filePath);
                Assert.fail("测试步骤有失败，整个测试用例执行失败");
                break;
            }
            if(testResult==true){
                log.info("测试用例执行结果为true");
                DataProviderFromExcel.setCellData(testSuiteNum, Constants.
                测试调度文件.suite_result, true, fileSheet, filePath);
                Assert.assertTrue(true, "测试用例执行成功");
            }
        }
    }
}

public void execute_Actions(int testcaseNum, String testCaseName) throws Exception{
    try{
        for(int i =0;i<method.length;i++){
            if(method[i].getName().equals(actionstep)){
                method[i].invoke(action, mobileElememnt,data);
                if(testResult==true){
                    log.info("测试步骤执行结果为true");
                    DataProviderFromExcel.setCellData(testcaseNum, Constants.
```

```
                    测试用例文件.Col_result, true, testCaseName, Constants.
                    测试excel文件.filepath);
                    break;
                }
                else{
                    log.info("测试步骤执行结果为false");
                    DataProviderFromExcel.setCellData(testcaseNum, Constants.
                    测试用例文件.Col_result, false, testCaseName, Constants.
                    测试excel文件.filepath);
                }
            }
        }
    }
}
```

创建 runAll.xml 文件，同时在 pom.xml 中指定 Maven 运行的 TestNG 配置文件为 runAll.xml。具体的配置参考 5.3.2 节。

在本次测试执行中，为了让执行结果失败，对于最后一项"I accept adds"的检查，故意设置为"yes"而不是"false"，所以整个测试结果为失败。查看控制台日志，如图 9-3 所示。

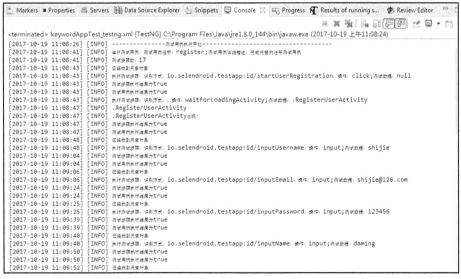

图 9-3　控制台日志

查看 test log.log 文件，如图 9-4 所示。

图 9-4　test log.log 文件

查看 ReportNG 测试报告，如图 9-5 所示。

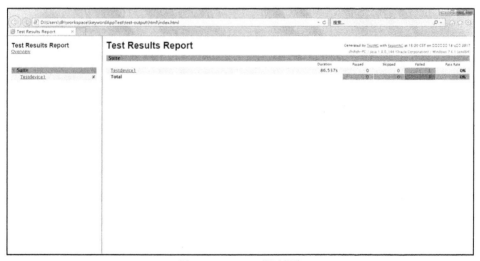

图 9-5　ReportNG 测试报告

查看数据源 Excel 执行日志。

（1）测试用例结果，如图 9-6 所示。

A	B	C	D
testCaseName	testCaseDetail	isRun	result
register	完成完整的注册测试用例	√	FALSE

图 9-6 测试用例结果

（2）测试步骤执行结果，如图 9-7 所示。

B	C	D	E	F	G	H
testStepID	testStepDetail	objectName	inspector	actionstep	data	result
testStep1	点击注册文件夹	首页-注册	io.selendroid.testapp:id/startUserRegi	click	null	TRUE
testStep2	等待页面跳转			waitForLoadingActivity	.RegisterUserActivity	TRUE
testStep3	输入用户名	注册页-用户名	io.selendroid.testapp:id/inputUsername	input	shijie	TRUE
testStep4	输入邮箱	注册页-邮箱	io.selendroid.testapp:id/inputEmail	input	shijie@126.com	TRUE
testStep5	输入密码	注册页-密码	io.selendroid.testapp:id/inputPassword	input	123456	TRUE
testStep6	输入姓名	注册页-姓名	io.selendroid.testapp:id/inputName	input	daming	TRUE
testStep7	点击选择语言	注册页-选择语言	io.selendroid.testapp:id/input_prefere	click	null	TRUE
testStep8	选择语言	注册页-语言	PHP	click	null	TRUE
testStep9	选择是否接受	注册页-是否接受	io.selendroid.testapp:id/input_adds	click_radio	yes	TRUE
testStep10	点击注册	注册页-点击注册	io.selendroid.testapp:id/btnRegisterUs	click	null	TRUE
testStep11	等待页面跳转			waitForLoadingActivity	.VerifyUserActivity	TRUE
testStep12	验证用户名	验证页-用户名	io.selendroid.testapp:id/label_usernam	verify	shijie	TRUE
testStep13	验证邮箱	验证页-邮箱	io.selendroid.testapp:id/label_email_d	verify	shijie@126.com	TRUE
testStep14	验证密码	验证页-密码	io.selendroid.testapp:id/label_passwor	verify	123456	TRUE
testStep15	验证姓名	验证页-姓名	io.selendroid.testapp:id/label_name_da	verify	daming	TRUE
testStep16	验证语言	验证页-语言	io.selendroid.testapp:id/label_prefere	verify	PHP	TRUE
testStep17	验证选择是否接受	验证页-是否接受	io.selendroid.testapp:id/label_acceptA	verify	yes	FALSE

图 9-7 测试步骤执行结果

使用关键字驱动测试框架，使得不懂技术的人可以实施自动化测试，便于自动化测试的推广。本测试框架只供参考，部分函数需要进行扩展和封装，用户可以根据项目实际情况重新修订。

第 10 章
持续集成的自动化

Martin Fowler 定义持续集成（Continuous Integration，CI）为一种软件开发实践，即团队开发成员经常集成它们的工作。每个成员每天至少集成一次，也就意味着每天可能会发生多次集成，每次集成都通过自动化的构建来验证，从而尽早地发现集成错误。在快速迭代的开发模式（如当下比较流行的敏捷研发模式）下，这种快速集成、快速自动化测试的模式是必需的。否则，传统的测试在这种迭代节奏下会显得捉襟见肘。

Appium 持续集成的过程为首先下载代码，然后编译代码，将代码转化为Json，再将Json命令发送给 Appium，最后 Appium 将代码发给手机进行执行。

10.1　安装 Jenkins

安装 Jenkins 非常简单，下载网址为 Jenkins 官网。截至本书编写时最新版本为 2.73.1。

在 Linux 操作系统下，可直接下载 war 包。如果中间件使用的是 Tomcat，直接放置在 webapps 目录下面，即可自动运行。或者在 Windows 命令行界面中手动运行"java‐jar Jenkins.war"也可以启动 Jenkins。

在 Windows 操作系统下，可下载 msi 版本，双击安装文件即可安装。

安装成功后，打开任意浏览器，在地址栏输入图 10-1 所示内容即可运行 Jenkins。

Jenkins 默认使用 8080 端口，如果该端口已经被占用，则可以在 Jenkins 安装目录 "/jenkins.xml" 中找到字段 "--httpPort=8080"，修改 8080 端口为指定的端口号。

首次打开 Jenkins 会要求输入密码，按照指定路径复制密码到编辑框即可，如图 10-1 所示。

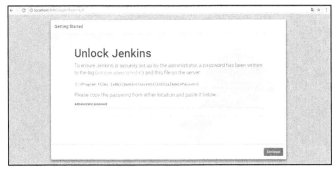

图 10-1　首次打开 Jenkins 时的密码输入界面

打开 Jenkins，进入欢迎界面，如图 10-2 所示。

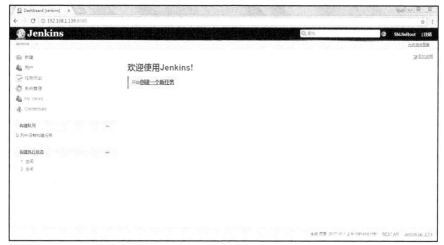

图 10-2　Jenkins 欢迎界面

10.1.1　安装 Jenkins 插件

Jenkins 集成了很多插件，凡是主流的开源软件都有集成。在主页面中，在左侧面板中选择"系统管理"，在右侧面板中选择"可用插件"标签。在过滤文本框中输入"maven"，选择与 Maven 相关的插件。除了 Maven 之外，需要安装的插件还包括 Git、TestNG Results Plugin 和 PowerShell plugin。以 Maven 插件安装为例，如图 10-3 所示，其他插件的安装方式类似，这里就不再详述了。

图 10-3　选择 Maven 插件

选择 Maven Integration plugin 进行安装，安装过程如图 10-4 所示。

图 10-4　安装 Maven 插件

插件安装完成之后，建议重启 Jenkins。

10.1.2　Jenkins 插件全局配置管理

下面配置这个项目中要用到的插件。首先从 SVN 或者 GitHub 中将代码下载下来。下载代码之前，首先要知道 JDK 的配置、Maven 的配置和 Git 的配置等，位置在"系统管理"→"Global Tool Configuration"。

1. JDK 插件配置

如图 10-5 所示，选择一个已经配置过的 JDK 版本，参考 3.1 节。

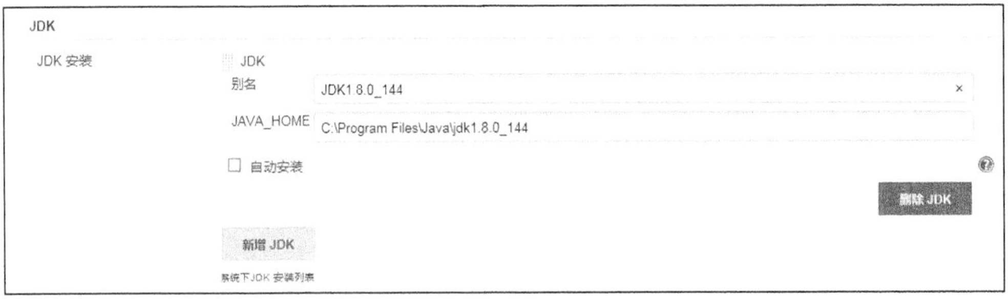

图 10-5　JDK 插件配置

2. Maven 配置

如图 10-6 所示，在 Jenkins 系统的环境变量中设置 Maven 变量。

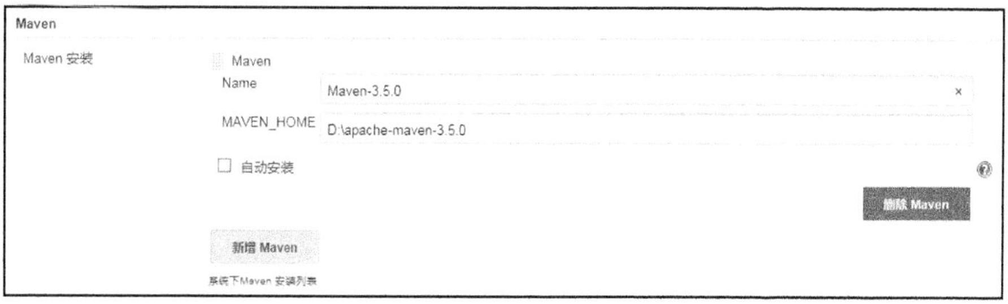

图 10-6　Maven 插件配置

- Name：指定 Maven 的名称，建议在名称后面添加 Maven 的版本号。因为在 Maven 项目构建时，需要指定一个具体的 Maven 版本号。
- MAVEN_HOME：指定 Maven 的具体路径，也就是 5.2.1 节中设置的 M2_HOME。

3. Git 插件配置

Git 用来帮助 Jenkins 从 GitHub 下载代码，如图 10-7 所示。

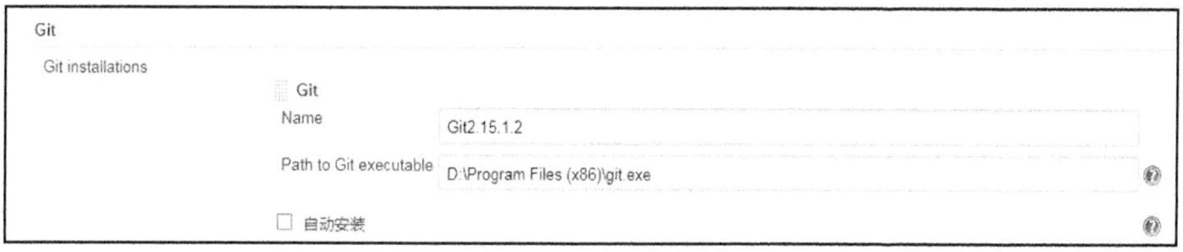

图 10-7　Git 配置

配置完后，直接保存即可。

10.2　Jenkins 持续集成基础配置

第 8 章和第 9 章中创建了两个 Maven 项目。本节以项目 AppTest 为例，实现使用 Jenkins 进行持续集成。

10.2.1　新建项目

在 Jenkins 主界面上单击"新建"按钮，在 Enter an item name 文本框中输入新的项目名

testRegister，选择"构建一个 maven 项目"，如图 10-8 所示。

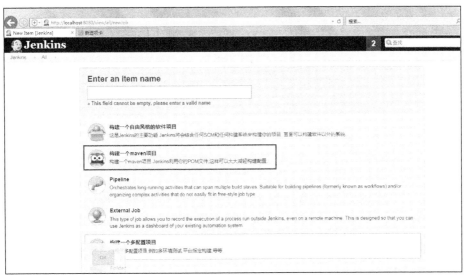

图 10-8　构建 Maven 项目

单击 OK 按钮，项目创建成功，同时跳转到项目配置页面，如图 10-9 所示。

图 10-9　项目配置页面的 General 选项卡

1．配置"源码管理"

"源码管理"选项卡，如图 10-10 所示。如果使用 SVN 或者 Git，输入项目所在路径的 URL，表示从 SVN 或者 Git 中指定路径下载对应的代码。

图 10-10　配置"源码管理"

2. 配置"构建触发器"

如何执行代码，是手动执行构建还是使用自动构建计划定期触发，勾选 Build whenever a SNAPSHOT dependency is built 复选框（见图 10-11），Jenkins 会解析 pom 文件，一般用于手动执行。

自动构建计划的设计和 Linux 的 Contrab 定时任务有点像，详细情况建议参考相关的说明文档。这里配置的含义为，每隔 15min 构建一次，如图 10-11 所示。

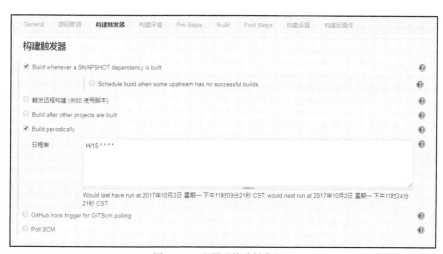

图 10-11　配置"构建触发器"

3. 配置 Pre Steps

对于一个 Maven 项目，最重要的当然是构建 Maven 项目。在构建 Maven 项目之前，可能需要做一些事情。在界面中单击 Add pre-build step 按钮，在弹出的下拉框中选择执行以下各种构建命令。

- Execute Windows Batch command：执行 Windows 批处理命令。
- Excute shell：执行 Shell 脚本。
- Invoke Ant：调用 Ant 脚本。

- Invoke Gradle script：调用 Gradle 脚本。
- Invoke top-level Maven targets：调用 Maven 项目等。

4．配置 Build

在 Build 页面中，配置以下选项（见图 10-12）。

- Root POM：用于设置指定的 POM 文件。
- Goal and options：用于设置待执行的 Maven 命令，本例中要执行"maven clean install"。这里填写"clean install"，也可以输入其他的 Maven 命令。
- MAVEN_OPTS：用于设置启动 Maven 时的 JVM 选项，也可以在 POM 文件中指定。在第 8 章中，在 pom.xml 文件中已经指定，这里不做设置。

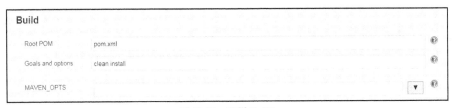

图 10-12　配置 Build

设置完成后，直接保存即可。

5．配置 Post Steps

有了前置步骤和构建步骤，自然少不了后置步骤。通过以下选项设置在什么情况下执行后置步骤。

- Run only if build succeeds：仅当构建成功时执行。
- Run only if build succeeds or is unstable：仅当构建成功或者不稳定时执行。
- Run regardless of build result：不管构建结果如何，直接执行。这里选中 Run regardless of build result 单选按钮，如图 10-13 所示。

图 10-13　选中 Run regardless of build result 单选按钮

同时，单击 Add post build step 下拉按钮，从下拉列表中选择 Execute shell。在 Command 文本框中，输入 call f:\maven.bat（见图 10-14）。

图 10-14　配置 Post Steps

放置在笔者计算机的 f 盘的 maven.bat 的内容如下。

```
cd %WORKSPACE%
D:\apache-maven-3.5.0\bin mvn test
```

在第 1 行中：转到下载 Jenkins 的 workspace 文件夹。

在第 2 行中：执行 mvn test。

在构建 Jenkins 时，会将项目下载到 Jenkins 安装目录"/workspace"下。

6．配置"构建后操作"

单击"增加构建后操作步骤"下拉按钮，选择 Publish TestNG Results，输出 TestNG 报告。需要注意的是，当采用 Git 或者 SVN 管理代码的时候，每次构建都会从代码库下载工作空间到 Jenkins 安装目录下的 workspace 文件夹中，报告也放置在对应的工作空间中，如图 10-15 所示。

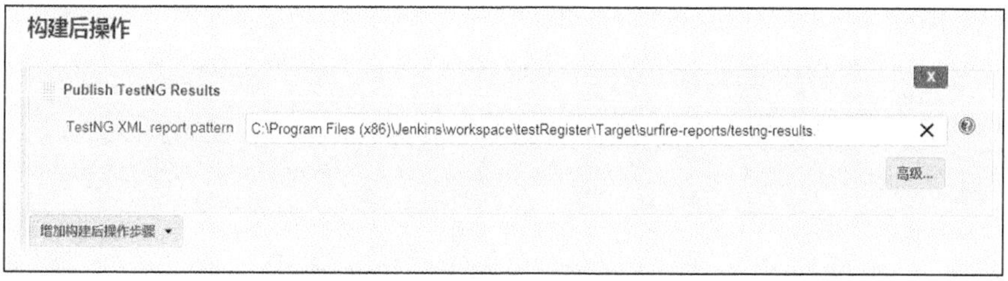

图 10-15　配置"构建后操作"

10.2.2 构建项目

配置完成后,单击"立即构建"选项,Jenkins 开始手动构建项目,如图 10-16 所示。

可以在控制台查看输出结果,如图 10-17 所示。

图 10-16 选择立即构建

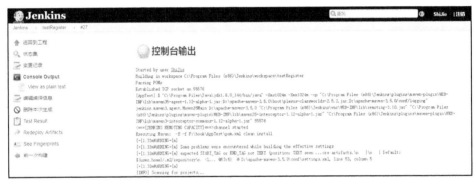

图 10-17 控制台输出

查看构建结果,如图 10-18 所示。

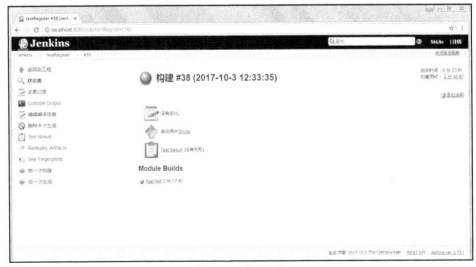

图 10-18 构建结果

查看测试结果，如图 10-19 所示。

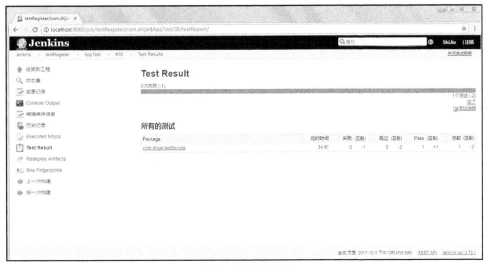

图 10-19　测试结果

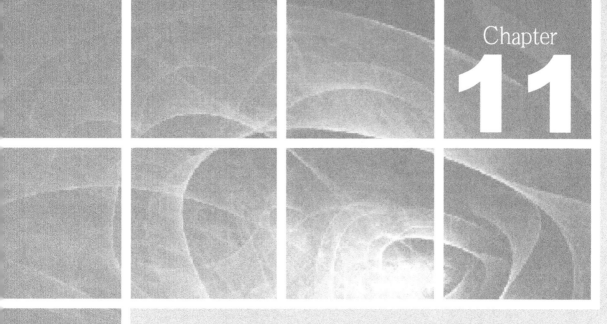

第 11 章

Appium 常见问题处理方式

本章对自动化测试中的常见问题进行总结，供平时遇到问题时进行参考。

11.1　输入中文

在使用 Appium 做手机端的自动化测试时，会遇到输入中文的问题。常见的解决方式如下。

（1）把测试类另存为 UTF-8 格式。

（2）在 Desired Capabilities 中增加两个属性：unicodeKeyboard 和 resetKeyboard。

```
capabilities.setCapability("unicodeKeyboard", true);
capabilities.setCapability("resetKeyboard", true);
```

在脚本的执行过程中可以直接输入中文，程序执行成功。

```
username.sendKeys("视界");
```

11.2　滑动操作

Appium 通过 swipe 函数处理滑动问题，如代码清单 11-1 所示。从代码可以看出，此方法共有 5 个参数，都是整型，依次是起始位置的 x、y 坐标和终点位置的 x、y 坐标，以及滑动间隔时间，单位为 ms。

代码清单 11-1　　　　　　　　　　swipe 函数

```
public void swipe(int startx, int starty, int endx, int endy, int duration) {
    TouchAction touchAction = new TouchAction(this);

    //Appium 把 press-wait-moveto-release 转换为滑动操作
    touchAction.press(startx, starty).waitAction(duration)
            .moveTo(endx, endy).release();

    touchAction.perform();
}
```

为了让 Appium 更好地兼容不同分辨率的设备，需要在滑动前先获取屏幕的分辨率。

```
int width = driver.manage().window().getSize().width;
int height = driver.manage().window().getSize().height;
```

然后根据屏幕的分辨率执行各种滑动操作。

```
//向上滑动
driver.swipe(width / 2, height * 3 / 4, width / 2, height / 4, during);
//向下滑动
driver.swipe(width / 2, height / 4, width / 2, height * 3 / 4, during);
//向左滑动
driver.swipe(width / 4, height / 2, width * 3 / 4, height / 2, during);
//向右滑动
driver.swipe(width * 3 / 4, height / 2, width / 4, height / 2, during);
```

以大众点评网的 App 为例，在大众点评网首页中，单击"美食"图标（如图 11-1 所示），弹出图 11-2 所示界面。

图 11-1 大众点评网首页

图 11-2 大众点评网的"美食"界面

"美食"界面罗列了很多美食的信息。"深×××"不在"美食"界面中，需要向上滑动页面，让"深×××"显示出来并单击进入"深×××"对应的列表项。

实现的代码如代码清单 11-2 所示。

代码清单 11-2　　　　　　　　　大众点评网示例

```
public void testSwip() throws InterruptedException{
    boolean up;
```

```java
//美食图标
MobileElement meishiElement = (MobileElement) driver.findElement(By.xpath
("// android.support.v7.widget.RecyclerView/android.widget.RelativeLayout[1]
/android.widget.ImageView"));
//单击美食图标，跳转到美食界面
meishiElement.click();
//获取当前页面的屏幕尺寸
int width = driver.manage().window().getSize().width;
int height = driver.manage().window().getSize().height;
while(true){
    try{
        MobileElement endElement = (MobileElement) driver.findElement(By.xpath
        ("//android.widget.TextView[@text='深×××']"));
        endElement.click();
        break;
    }catch(Exception e){
        driver.swipe(width / 2, height * 3 / 4, width / 2, height / 4, 1000);
        Thread.sleep(2000);
    }
}
```

11.3 滚动操作

Appium 早期版本提供了 Scroll 方法来实现滚动，最新的版本已经取消了这个方法。笔者使用 findElementByAndroidUIAutomator()方法来实现同样的功能，如代码清单 11-3 所示。

代码清单 11-3　　　　　　　　滚动操作的实现

```java
public void scrollToElement(String str) {
    ((AndroidDriver<MobileElement>) driver).findElementByAndroidUIAutomator(
        "new UiScrollable(new UiSelector().scrollable(true).instance(0)).ScrollIntoView
        (new UiSelector().textContains(\""+ str + "\").instance(0))");
}
```

```java
public void scrollToExactElement(String str) {
    ((AndroidDriver<MobileElement>) driver).findElementByAndroidUIAutomator(
            "new UiScrollable(new UiSelector().scrollable(true).instance(0)).scrollIntoView"
            + "(new UiSelector().textContains(\""+ str + "\").instance(0))");
}
```

findElementByAndroidUIAutomator()要求传入的参数为字符串。

UiSelector().scrollable(true) .instance(0)表示找到一个可以滑动的对象，通过判断 scrollable 属性是否为 true 进行查找。

scrollIntoView 滑动到匹配的 selector 控件，如果没有匹配到，则停留在滑动列表的最下方。

在 new UiSelector().text(string).instance(0)待匹配的 selector 控件中，设置 text 属性值为指定的 string。

UiSelector().textContains(string).instance(0))待匹配的 selector 控件，指定 text 属性值包含指定的 string。

11.4 输入 Android 按键

在清除编辑框的内容时，采用以下步骤。

（1）获取编辑框中文本的长度。

（2）将光标移动到文本的尾部。

（3）按退格直到所有文本被删除。

如代码清单 11-4 所示，参考 6.2.2 节，按键 123 表示光标移动到末尾，按键 67 表示退格键。

代码清单 11-4　　　　　　　　　　输入 Android 按键

```java
/**
 * 描述：清理文本框
 * 参数：文本框的内容
 */
public void clearText(String text) {
    for (int i = 0; i < text.length(); i++) {
```

```
//123: KEYCODE_MOVE_END 光标移动到末尾
            driver.pressKeyCode(123);
//KEYCODE_DEL 退格键 67
            driver.pressKeyCode(67);
        }
    }
}
```

11.5　处理 Popup Window

Popup Window 是一个弹出窗口控件，可以用来显示任意视图（View），而且会浮动在当前活动（Activity）的顶部。通过 UI Automator Viewer 无法识别，通过 Hierarchy Viewer 才可以识别到 Popup Window。这属于系统逻辑，Appium 暂时没有处理机制，可以采用下面的方法进行解决。

- 通过 Tap 方法，代码如下。

```
WebElement imagebtn = driver.findElementById("showPopupWindowButton");
magebtn.click();
driver.tap(1, 214, 475, 10);
```

- 通过 TouchAction，代码如下。

```
WebElement imagebtn2 = driver.findElementById("showPopupWindowButton");
imagebtn2.click();
TouchAction action = new TouchAction(driver);
action.press(214, 475).release().perform();
```

对于需要操作的 Popup Window，用户可以首先获得坐标点，然后以封装为 Hash 函数的方式进行操作。例如弹出框上有 4 个蓝色的球，要获取每个球的坐标点，可以封装这 4 个坐标点到一个 Hash 函数中，如代码清单 11-5 所示，在其他方法中通过调用 Hash 函数实现对任意蓝色的球进行操作，如代码清单 11-6 所示。这里采用的是绝对坐标，考虑到分辨率的问题，建议对相对坐标进行封装。

代码清单 11-5　　　　　　　　　　获取坐标点并封装

```
package appiumsample;
import java.util.*;
```

```java
import org.openqa.selenium.Point;
public class Popuppointer {
    //获得蓝球的坐标
    public void blueball(){
    Point x1 = new Point(1013, 534);
    Point x2 = new Point(1114,534);
    Point x3 = new Point(1210,534);
    Point x4 = new Point(1294,534);
    HashMap<Integer, Point> blueball= new HashMap<Integer, Point>();
    blueball.put(1,x1);
    blueball.put(2,x2);
    blueball.put(3,x3);
    blueball.put(4,x4);
    }
}
```

调用方法如代码清单 11-6 所示。

代码清单 11-6　　　　　　　　调用 Hash 函数

```
driver.tap(1,Popuppointer.blueball().get(1).x,Popuppointer.blueball().get(1).y,30);

driver.tap(1,Popuppointer.blueball().get(2).x,Popuppointer.blueball().get(2).y,30);

driver.tap(1,Popuppointer.blueball().get(3).x,Popuppointer.blueball().get(3).y,30);

driver.tap(1,Popuppointer.blueball().get(4).x,Popuppointer.blueball().get(4).y,30);
```

Popup Window 无法识别，很大程度上是因为焦点无法找到，以上方法仅适用于无源码的情况。

11.6　处理 Toast

Toast 是 Android 中用来显示信息的一种机制，和 Dialog 不一样的是，Toast 是没有焦点的，而且 Toast 显示的时间有限，过一定的时间就会自动消失。此外，它主要用于向用户显示提示消息。Toast 属于 Android 系统逻辑，不属于应用逻辑，所以通过 UI Automator Viewer 无法获取控件。可以采用下面的方法进行解决。

1. 图片比对方法

在 Toast 被触发后，截取界面的图片，然后进行比对，如代码清单 11-7 所示。

代码清单 11-7　　　　　　　　　　　对比图片

```
WebElement imagebtn = driver.findElementById("showToastButton");
imagebtn.click();
try{
    File scrFile =driver.getScreenshotAs(OutputType.FILE);
    FileUtils.copyFile(scrFile, new File("D:\\scrshot.png"));
}
catch(Exception e){
    e.printStackTrace();
}
```

通过 getScreenshotAs 方法来捕捉屏幕，如代码清单 11-7 所示，使用 OutputType.FILE 作为参数传递给 getScreenshotAs，告诉它将截取的屏幕以文件的形式返回。使用 copyFile 保存 getScreenshotAs 截取的屏幕文件到 D 盘中并命名为 scrshot.png。

 因为 Toast 一般显示有限的时间就会自动消失，所以在截取图片的时候，建议多截取几张，以保证截取成功。

2. Seledroid 方法

Seledroid 方法（自动化测试引擎）可以识别 Toast 控件，采用 WaitForElement 方法获得 Toast 上的文本。

```
waitForElement(By.partialLinkText("Hello seledroid toast"), 4, driver);
```

3. 使用 Automator2

在最新版本的 Appium 中，使用 Automator2 自动化测试引擎，可以获取 Toast。升级 Appium 为 GUI 1.2.3 版本，如代码清单 11-8 所示。

代码清单 11-8　　　　　　　　在 Appium 中处理 Toast

```
package com.shijie.testScripts;
```

```java
import static org.testng.Assert.assertNotNull;
import io.appium.java_client.android.Activity;
import io.appium.java_client.android.AndroidDriver;
import java.io.File;
import java.net.URL;
import java.util.concurrent.TimeUnit;
import org.openqa.selenium.By;
import org.openqa.selenium.WebElement;
import org.openqa.selenium.remote.DesiredCapabilities;
import org.openqa.selenium.support.ui.ExpectedConditions;
import org.openqa.selenium.support.ui.WebDriverWait;
import org.testng.annotations.AfterMethod;
import org.testng.annotations.BeforeMethod;
import org.testng.annotations.Test;
import io.appium.java_client.remote.AutomationName;
import io.appium.java_client.remote.MobileCapabilityType;

public class testkongjian {
    AndroidDriver<WebElement> driver;
    @BeforeMethod
    public void SetUp() throws Exception {
        File appDir = new File("F:\\");
        File app = new File(appDir, "selendroid-test-app-0.17.0.apk");
        DesiredCapabilities capabilities = new DesiredCapabilities();
        capabilities.setCapability(MobileCapabilityType.DEVICE_NAME, "Android Emulator");
        capabilities.setCapability(MobileCapabilityType.APP, app.getAbsolutePath());
        capabilities.setCapability(MobileCapabilityType.AUTOMATION_NAME,
          AutomationName.ANDROID_UIAUTOMATOR2);
        driver = new AndroidDriver<>(new URL("http://127.0.0.1:4723/wd/hub"), capabilities);
        driver.manage().timeouts().implicitlyWait(30, TimeUnit.SECONDS);
    }


    @Test
    public void test_toast(){
        Activity activity = new Activity("io.selendroid.testapp", ".HomeScreenActivity");
```

```
    driver.startActivity(activity);
    WebElement toastButton = null;
    toastButton = driver.findElement(By.id("io.selendroid.testapp:id/showToastButton"));
    toastButton.click();
     final WebDriverWait wait = new WebDriverWait(driver, 10);
    assertNotNull(wait.until(ExpectedConditions.presenceOfElementLocated(
        By.xpath("//*[@text='Hello selendroid toast!']"))));
}

@AfterMethod
public void TearDown()
{
    driver.quit();
}

/**
 * @param args
 */
public static void main(String[] args) {
    // TODO Auto-generated method stub
}
}
```

11.7 处理长按

长按是用户比较常用的一种手势,与点触后一直持续按下的动作不同,有时候长按可以获得更多的信息提醒。可以使用 TouchAction 模拟长按操作,如在拨号栏中长按*会显示值"P",如代码清单 11-9 所示。

操作步骤如下。

(1)创建 TouchAction 对象。

(2)执行 Longpress 操作。

(3)使用 perform 将操作名传递给手机 App。

代码清单 11-9　　　　　　　　　　　　　长按

```java
private AndroidDriver driver;
    @Before
    public void setUp() throws Exception {
        DesiredCapabilities capabilities = new DesiredCapabilities();
        capabilities.setCapability("deviceName", "Lenovo A788t");//真机名称为 Lenovo A788t
        capabilities.setCapability("platformVersion", "4.3");//操作系统版本为 4.3
        capabilities.setCapability("platformName", "Android");//操作系统名称为 Android
        capabilities.setCapability("udid", "00a10399");//使用的真机为 Android 平台
        capabilities.setCapability("appPackage", "com.android.contacts");//待测 App 包
        capabilities.setCapability("appActivity",".activities.DialtactsActivity");
        //待测 App 主 activity 名
        capabilities.setCapability("automationName", "appium");//
        driver = new AndroidDriver(new URL("http://127.0.0.1:4723/wd/hub"),capabilities);
    }
    @Test
    public void testlongpress(){
        WebElement dail = driver.findElementByName("拨号");
        dail.click();
        TouchAction taction =new TouchAction(driver);
        WebElement star = driver.findElementById("star");
        taction.longPress(star).perform();
        WebElement result = driver.findElementById("digits");
        assert result.getText().equals("P"):"Actual value is"+result.getText()+"did not match expected value: P";
    }
    @After
    public void stop() {
        driver.quit();
    }
}
```

11.8　处理下拉列表框

Selenium 针对 WebList 下拉列表框采用了 Select 类，但是 Appium 并未提供该类的处理，而且由于 Appium 底层 UIAutomator 的限制，超过屏幕界限的下拉选项未被选择进来。这里

提供对 WebList 类的处理，如在 selendroid-test-app-0.15.0.apk 中选择编程语言。

对于图 11-3 所示的单选按钮，class 都为 CheckedTextView。

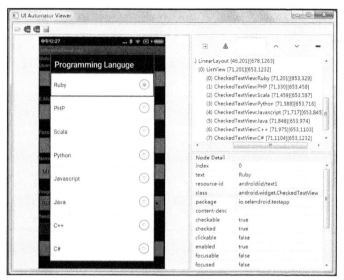

图 11-3　查看 CheckedTextView 类

处理方式如代码清单 11-10 所示。

代码清单 11-10　　　　　　　　　处理下拉列表框

```
public void checkedTextView(){
    //使用class属性选择所有的单选按钮,并存放在一个list中
    @SuppressWarnings("unchecked")
    List<MobileElement> checkedTextViews = (List<MobileElement>) driver.
    findElements ByClassName("android.widget.CheckedTextView");
    //使用for循环遍历list中的每个单选按钮,查到name值为Ruby的单选按钮,如果该单选按钮未处于
    //选中状态,则调用click方法进行选择
    for(MobileElement checkedTextView:checkedTextViews){
        if(checkedTextView.getAttribute("name").equals("Ruby")){
            if(!checkedTextView.isSelected()){
                checkedTextView.click();}
            Assert.assertTrue(checkedTextView.isSelected());
        }
    }
}
```

11.9　处理缩放

缩放手势类似于拖动手势，第二个手指并拢之后按下，接着两个手指向相同或者相反的方向滑动，从而实现页面的放大或者缩小。Appium 支持多点触摸，可以使用 MultiTouchAction 类同时启用多个 TouchAction 类，如将相册中的图片放大。

实现方式为通过 TouchAction 封装多个 touch 动作，并把 Touch 动作封装到 MultiTouchAction 中。在 zoom() 方法中 x、y 表示偏移量，如代码清单 11-11 所示。

代码清单 11-11　　　　　　　　　　缩放操作

```java
public void zoom(int x, int y) {
    MultiTouchAction multiTouch = new MultiTouchAction(driver);
    int scrHeight = driver.manage().window().getSize().getHeight();
    int yOffset = 100;
    if (y - 100 < 0) {
      yOffset = y;
    } else if (y + 100 > scrHeight) {
      yOffset = scrHeight - y;
    }
    TouchAction action0 = new TouchAction(driver).press(x, y).moveTo(x, y - yOffset).release();
    TouchAction action1 = new TouchAction(driver).press(x, y).moveTo(x, y + yOffset).release();
    multiTouch.add(action0).add(action1);
    multiTouch.perform();
}
```

11.10　检查元素文本是否可见

以 selendroid-test-app-0.15.0.apk 为例，当单击 Display text view 按钮时才会出现 "Text is sometimes displayed" 所在的控件，如图 11-4 所示。

判断文本是否出现的代码如代码清单 11-12 所示。

11.10 检查元素文本是否可见 | 349

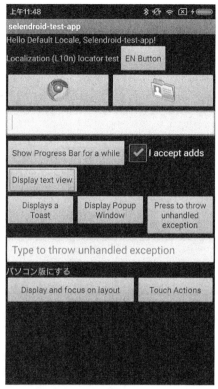

图 11-4 单击 Display text view 按钮

代码清单 11-12   检查元素文本是否为可见

```
@Test
public void testElementPresent() throws InterruptedException{
    String expectValue = "Text is sometimes displayed";
    //识别控件 Display text view
    MobileElement visibleButtonElement = (MobileElement) driver.findElementById
    ("io.selendroid.testapp:id/visibleButtonTest");
    //单击 Display text view 按钮
    visibleButtonElement.click();
    //弹出文本通过 By.id 进行识别
    By visibileButton_by = By.id("io.selendroid.testapp:id/visibleTextView");
    //判断弹出的文本是否存在
    if(isElementPresented(visibileButton_by)){
        //获取文本内容
        String visibileButtonText = driver.findElement(visibileButton_by).getAttribute ("text");
```

```java
            //判断文本内容是否包含预期值。若包含则打印成功信息；若不包含，则用例执行失败
            if(visibileButtonText.contains(expectValue)){
                System.out.println("查找控件成功");
            }
            else{
                Assert.fail("查找控件失败");
            }
        }
    }

    public boolean isElementPresented(By by){
        boolean isDisplayed = false;

        try{
            isDisplayed = driver.findElement(by).isDisplayed();
        }catch(NoSuchElementException e){
            isDisplayed = false;
        }
        return isDisplayed;
    }
```

值得注意的是，如果 driver.findElement(by)无法查找到控件元素，将会报出异常，后面的 isDisplayed()方法不会被执行，所以需要增加 try catch 以捕获异常。

11.11　启动其他 App

在测试中，也许会涉及多个 App 之间的交互，或者在一个 App 中会启动其他 App。如在测试大众点评 APK 时，在首页中单击"美食"图标进入"美食"界面，这时启动 selendroid-test-app-0.17.0.apk。实现多 APK 交互的代码如代码清单 11-13 所示。

代码清单 11-13　　　　　　　　多 APK 交互

```java
@Test(description="测试启动 app")
public void TeststartActivity(){
    //"美食"图标
```

```
    MobileElement meishiElement = (MobileElement) driver.findElement(By.xpath ("//android.
    support.v7.widget.RecyclerView/android.widget.RelativeLayout[1]/android.widget.
    ImageView"));
    //单击"美食"图标，跳转到"美食"界面
    meishiElement.click();
    //启动 selendroid-test-app 的包名
    String androidPackage = "io.selendroid.testapp";
    //启动 selendroid-test-app 的 main Activity
    String androidStartActivity = ".HomeScreenActivity";
    //启动 selendroid-test-app
    driver.startActivity(new Activity(androidPackage, androidStartActivity));
}
```

11.12 并行测试

并行（多线程）技术在软件术语里被定义为软件、操作系统或程序可以并行地执行另外一段程序中多个部分或者子组件的能力。并行（多线程）测试可以给用户带来很多好处，主要包括以下两点。

- 缩短了执行时间。并行测试也就意味着多个测试可以同时执行，从而减少了整体测试所花费的时间。
- 允许多个线程并行地测试同一个测试组件。有了这个特性，用户就能够写出相应的测试用例来验证应用程序中包含多线程部分的代码的正确性。

Appium 提供了一种方式以在一台设备上自动操作多个 Android 会话，即采用多个标识符的方式启动多个 Appium 服务器端。

本节以在一台设备上运行两个 Appium 服务器为例，讲解如何进行并行测试。具体步骤如下。

（1）创建 Maven 项目。

（2）启动两个 Appium 服务器，并配置端口号。

对于第一个 Appium 服务器：Server Port 为 4723，Bootstrap Port 为 4724。

对于第二个 Appium 服务器：Server Port 为 4725，Bootstrap Port 为 4726。

在 Appium 的 GUI 中，选择 Advanced 标签，在 General 选项组下，在 Server Port 文本框中设置服务器端口号，如图 11-5 所示。

图 11-5 Appium 服务器端口号设置

选择 Advanced 标签，在 General 选项组下，在 Bootstrap 文本框中设置 Bootstrap 端口号，如图 11-6 所示。

图 11-6 Bootstrap 端口号设置

下面以 selendroid-test-app-0.17.0.apk 为例，介绍如何在两个手机上同时运行测试用例。

（1）创建 Constants 用来存储手机配置信息，如代码清单 11-14 所示。

代码清单 11-14　　　　　　　　　　Constants

```
package com.shijie.util;
public class Constants {
    public class 测试手机_红米{
        public static final String deviceName = "Redmi 3";
        public static final String udid = "c1aeae297d72";
        public static final String platformVersion = "5.1.1";
        public static final String platformName = "Android";
        public static final String appPackage = "io.selendroid.testapp";
        public static final String appActivity = ".HomeScreenActivity";
        public static final String unicodeKeyboard = "True";
        public static final String noSign = "True";
    }
    public class 测试手机_小米{
        public static final String deviceName = "2014813";
        public static final String udid = "59de0f3";
        public static final String platformVersion = "4.4.4";
        public static final String platformName = "Android";
        public static final String appPackage = "io.selendroid.testapp";
        public static final String appActivity = ".HomeScreenActivity";
        public static final String unicodeKeyboard = "True";
        public static final String noSign = "True";
    }
}
```

（2）创建两个 testNG 测试用例。测试用例 1 如代码清单 11-15 所示，它用于完成注册。

代码清单 11-15　　　　　　　　　　测试用例 1

```
package com.shijie.testScripts;
import java.net.URL;
import io.appium.java_client.android.AndroidDriver;
import io.appium.java_client.android.AndroidElement;
```

```java
import org.junit.Assert;
import com.shijie.util.Constants;
import org.openqa.selenium.WebElement;
import org.openqa.selenium.remote.DesiredCapabilities;
import org.testng.annotations.BeforeMethod;
import org.testng.annotations.AfterMethod;
import org.testng.annotations.Test;
public class testRegister{
    AndroidDriver<AndroidElement> driver;
    @BeforeMethod
@Parameters({"device_ID","port"})
    public void SetUp() throws Exception  {
        // TODO Auto-generated method stub
        capabilities.setCapability("deviceName", Constants.测试手机_红米.deviceName);
        capabilities.setCapability("udid", Constants.测试手机_红米.udid);
        capabilities.setCapability("platformVersion", Constants.测试手机_红米.platformVersion);
        capabilities.setCapability("platformName", Constants.测试手机_红米.platformName);
        capabilities.setCapability("appPackage", Constants.测试手机_红米.appPackage);
        capabilities.setCapability("appActivity", Constants.测试手机_红米.appActivity);
        capabilities.setCapability("unicodeKeyboard", Constants.测试手机_红米.unicodeKeyboard);
        capabilities.setCapability("noSign", Constants.测试手机_红米.noSign);
        driver = new AndroidDriver<>(new URL("http://"+device_ID+":"+port+"/wd/hub"), capabilities);}
    @Test
    public void testWebApp(){
        //对象识别
        WebElement startRegister_btn = driver.findElementById("io.selendroid.testapp:id/startUserRegistration");
        //单击,页面跳转
        startRegister_btn.click();
        //用户名对象
        WebElement username_txt = driver.findElementById("io.selendroid.testapp:id/inputUsername");
        username_txt.sendKeys("shijie");
        //E-mail 对象
```

```java
        WebElement email_txt = driver.findElementById("io.selendroid.testapp:id/inputEmail");
        email_txt.sendKeys("shijie@126.com");
        //密码对象
        WebElement password_txt = driver.findElementById("io.selendroid.testapp:id/inputPassword");
        password_txt.sendKeys("123456");
        WebElement name_txt = driver.findElementById("io.selendroid.testapp:id/inputName");
        name_txt.clear();
        name_txt.sendKeys("daming");
        //driver.hideKeyboard();
        WebElement languge_sel = driver.findElementById("io.selendroid.testapp:id/input_preferedProgrammingLanguage");
        languge_sel.click();
        WebElement prgLanguage = driver.findElementByName("Scala");
        prgLanguge.click();
        WebElement accept_check = driver.findElementById("io.selendroid.testapp:id/ input_adds");
        accept_check.click();
        WebElement register_btn = driver.findElementById("io.selendroid.testapp:id/btnRegisterUser");
        register_btn.click();
        WebElement label_name_data = driver.findElementById("label_name_data");
        Assert.assertEquals(label_name_data.getText().toString(), "daming");
    }
    @AfterMethod
    public void TearDown()
    {
        driver.quit();
    }
}
```

测试用例 2 如代码清单 11-16 所示，在图 11-4 所示主页面中单击 Display text view 按钮，验证弹出的文字是否正确。

代码清单 11-16　　　　　　　　测试用例 2

```java
package com.shijie.testScripts;
import java.net.URL;
import io.appium.java_client.android.AndroidDriver;
```

```java
import io.appium.java_client.android.AndroidElement;
import org.junit.Assert;
import org.openqa.selenium.NoSuchElementException;
import org.openqa.selenium.WebElement;
import com.shijie.util.Constants;
import org.openqa.selenium.remote.DesiredCapabilities;
import org.testng.annotations.BeforeMethod;
import org.testng.annotations.AfterMethod;
import org.testng.annotations.Parameters;
import org.testng.annotations.Test;
import org.openqa.selenium.By;

public class testDisplaytextView{
    AndroidDriver<AndroidElement> driver;
    @BeforeMethod
    @Parameters({"device_ID","port"})
    public void SetUp(String device_ID, String port ) throws Exception  {
            DesiredCapabilities capabilities = new DesiredCapabilities();
            capabilities.setCapability("deviceName", Constants.测试手机_小米.deviceName);
            capabilities.setCapability("udid", Constants.测试手机_小米.udid);
            capabilities.setCapability("platformVersion", Constants.测试手机_小米.platformVersion);
            capabilities.setCapability("platformName", Constants.测试手机_小米.platformName);
            capabilities.setCapability("appPackage", Constants.测试手机_小米.appPackage);
            capabilities.setCapability("appActivity", Constants.测试手机_小米.appActivity);
            capabilities.setCapability("unicodeKeyboard", Constants.测试手机_小米.unicodeKeyboard);
            capabilities.setCapability("noSign", Constants.测试手机_小米.noSign);
            driver = new AndroidDriver<>(new URL("http://"+device_ID+":"+port+"/wd/hub"),capabilities);
        }
    @Test
    public void testWebApp(){
        //对象识别
        boolean flag = false;
        String actualTxt = null;
        WebElement visibleButtonTest_btn = driver.findElementById("io.selendroid.testapp:id/visibleButtonTest");
        //单击,弹出页面文字
```

```java
        visibleButtonTest_btn.click();
        //判断文字是否弹出
        flag = isElementExist(By.id("visibleTextView"));
        if(flag){
            WebElement visibleText_text = driver.findElementById("io.selendroid.testapp:id/visibleTextView");
            actualTxt = visibleText_text.getAttribute("text");
            //判断文字是否一致
            Assert.assertEquals(actualTxt, "Text is sometimes displayed");
        }
        else{
            //文字没有弹出，断言失败
            Assert.fail("文字未弹出，测试用例失败");
        }
    }

    public boolean isElementExist(By locator){
        boolean flag = false;

        try{
            driver.findElement(locator);
            flag = true;
        }catch(NoSuchElementException e){

        }
        return flag;
    }
    @AfterMethod
    public void TearDown()
    {
        driver.quit();
    }
}
```

为了演示实例，以上操作只提取了 port、udid、devicename 作为参数，在实际操作中为了更好地实现参数与脚本分离，可以将所有的参数都提取出来。

(3）设置 testng.xml，如代码清单 11-17 所示。

代码清单 11-17　　　　　　　　　　　设置 testng.xml

```xml
<?xml version="1.0" encoding="UTF-8"?>
<!DOCTYPE suite SYSTEM "http://testng.org/testng-1.0.dtd">
<suite name="Suite" parallel="tests" thread-count="2" >
    <test name="Testdevice1" >
        <parameter name ="device_ID" value = "127.0.0.1"/>
        <parameter name ="port" value = "4723"/>
        <classes>
        <class name="com.shijie.testScripts.testRegister" />
        </classes>
    </test><!--Test-->
        <test name="Testdevice2" >
        <parameter name ="device_ID" value = "127.0.0.1"/>
        <parameter name ="port" value = "4725"/>
        <classes>
        <class name="com.shijie.testScripts.testDisplaytextviewClick" />
        </classes>
    </test><!--Test-->
</suite><!--Suite-->
```

在 testng.xml 上右击，就可以看到测试用例在两个手机上一起运行。

11.13　处理拖动

拖动就是将一个对象从一个位置拖到另外一个位置，可以简化桌面操作，如代码清单 11-18 所示。

代码清单 11-18　　　　　　　　　　　拖动

```java
public void drag(By startElement_by, By endElement_by){
    TouchAction act = new TouchAction(driver) ;
    //定位元素的原位置
    MobileElement startElement = (MobileElement) driver.findElement(startElement_by);
```

```
        //定位元素要移动到的目标位置
        MobileElement endElement = (MobileElement) driver.findElement(endElement_by) ;
        //执行元素的移动操作
        act.press(startElement).perform();
        act.moveTo(endElement).release().perform();
}
```

11.14 处理截图

Appium 可以通过使用 getScreenshotAs 截取整个页面作为图片,在测试过程中帮助我们直观地定位错误,如代码清单 11-19 所示。

代码清单 11-19　　　　　　　　截图操作

```
WebElement RegisterPage=driver.findElement(By.name("startUserRegistration"));
File screenShot=driver.getScreenshotAs(OutputType.FILE);
File location=new File("screenshots");
String screenShotName=location.getAbsolutePath()+File.separator+"testCalculator.png";
        try{
            System.out.println("save screenshop");
            FileUtils.copyFile(screenShot,new File(screenShotName));}
        catch(IOException e){
            System.out.println("save screenshop fail");
            e.printStackTrace();
        }
        finally{
            System.out.println("save screenshop finish");
        }
```

受到设备存储容量的限制,我们可以考虑扩展这个功能,使得它可以截取页面上某一个元素。要截取页面上的 username 编辑框,代码如代码清单 11-20 所示。

代码清单 11-20　　　　　　　　截取指定元素

```
WebElement username = driver.findElementById("inputUsername");
username.sendKeys("bree");
getElementShotSaveAs(username);
Assert.assertEquals("liming", username.getText());
```

```java
public void getElementShotSaveAs(WebElement element) throws IOException{
        File screenShot=driver.getScreenshotAs(OutputType.FILE);
        BufferedImage img = ImageIO.read(screenShot);
        int width = element.getSize().width;
        int height = element.getSize().height;
        Rectangle rect = new Rectangle(width,height);
        Point p = element.getLocation();
        BufferedImage dest = img.getSubimage(p.x, p.y, rect.width, rect.height);
        ImageIO.write(dest, "png",screenShot);
}
```

由于自动化测试是无人值守的，因此可以利用 TestNG 监听器来实现监听功能。当测试处于某种状态的时候执行错误截图，如测试失败时的截图。这里采用 testListenerAdapter 方法，每次测试失败的时候，都会重写该方法。

新建两个类，一个用作监听器，另外一个用于写测试代码。

1. 监听器

监听器是一些预定义的 Java 接口。用户创建这些接口的实现类，并把它们加入 TestNG 中，TestNG 便会在测试运行的不同时刻调用这些类中的接口方法。这里使用 ITestListener 监听器，实现其方法为 onTestFailure 在测试失败的时候，保存控件的截图，如代码清单 11-21 所示。

代码清单 11-21　　　　　　　　　　　监听器

```java
package appiumsample;
import java.io.File;
import java.io.IOException;
import io.appium.java_client.AppiumDriver;
import org.apache.commons.io.FileUtils;
import org.openqa.selenium.OutputType;
import org.testng.ITestResult;
import org.testng.TestListenerAdapter;

public class ScreenshotListener extends  TestListenerAdapter {
    @Override
    public void onTestFailure(ITestResult tr){
            AppiumDriver driver=Screenshot.getDriver();
```

```
            File location=new File("screenshots");
            String screenShotName=location.getAbsolutePath()+File.separator+tr.getMethod().getMethodName()+".png";
            //使用ItestResult获取失败方法名
            File screenShot=driver.getScreenshotAs(OutputType.FILE);
            try {
            FileUtils.copyFile(screenShot,new File(screenShotName));
            }
            catch(IOExceptione){
            e.printStackTrace();
            }
        }
    }
}
```

2. 测试代码

通过使用"@Listeners"注释，可以直接在 Java 源代码中添加 TestNG 监听器，如代码清单 11-22 所示。

代码清单 11-22　　　　　　　　　测试代码

```
package appiumsample;
import io.appium.java_client.android.AndroidDriver;
import java.io.IOException;
import java.net.MalformedURLException;
import java.net.URL;
import org.openqa.selenium.WebElement;
import org.openqa.selenium.remote.DesiredCapabilities;
import org.testng.Assert;
import org.testng.annotations.AfterClass;
import org.testng.annotations.BeforeClass;
import org.testng.annotations.Listeners;
import org.testng.annotations.Test;


@Listeners({ScreenshotListener.class})
public class Screenshot {
    private static AndroidDriver driver;
    @BeforeClass
```

```java
    public void setup() throws MalformedURLException {
        //App 地址
        String apppath = "F:\\selendroid-test-app-0.15.0_debug.apk";
        //配置 AndroidDriver
        DesiredCapabilities capabilities = new DesiredCapabilities();
        capabilities.setCapability("deviceName", "Lenovo A788t");//真机的名称为 Lenovo A788t
        capabilities.setCapability("platformVersion", "4.3");//操作系统版本为 4.3
        capabilities.setCapability("platformName", "Android");//操作系统名称为 Android
        capabilities.setCapability("udid", "00a10399");//使用的真机为 Android 平台
        capabilities.setCapability("app", apppath);//确定待测 App
        capabilities.setCapability("appPackage", "io.selendroid.testapp");//待测 App 包
        capabilities.setCapability("appActivity", ".HomeScreenActivity");//待测 App 主 Activity 名
        capabilities.setCapability("automationName", "selendroid");
        driver = new AndroidDriver(new URL("http://127.0.0.1:4723/wd/hub"),capabilities);
        // WebDriverWait wait = new WebDriverWait(driver,10);
    }
    @SuppressWarnings("deprecation")
@Test
    public void testExample() throws IOException {
        WebElement username = driver.findElementById("inputUsername");
        username.sendKeys("bree");
        Assert.assertEquals("liming", username.getText());
    }
    public static AndroidDriver getDriver(){
        return driver;
    }
    @AfterClass
    public void tearDown(){
    }
}
```

11.15 隐式等待

在运行测试时,测试可能并不总是以相同的速度响应,例如,可能在几秒后进度条到

100%时,按钮才会变成可单击的状态。这里介绍不同的方法进行同步测试。

隐式等待有两种方法,即 implicitlyWait 和 sleep。需要注意的是,一旦设置了隐式等待,则它存在整个 driver 对象实例的生命周期中。在下例中,设置全局等待时间是 30s,这是最长的等待时间。

最直接的方式是设置固定的等待时间。

```
Thread.sleep(30000)
```

对于固定等待时间的元素,可以用 sleep 进行简单的封装来实现等待指定的时间,如代码清单 11-23 所示。

代码清单 11-23   用 sleep 实现等待

```java
@Test(description = "sleep 简单封装")
    private boolean testisElementPresent(By by) throws InterruptedException {
        try {
        //设置等待时间
          Thread.sleep(1000);
          //查找元素
          driver.findElement(by);
          //若找到元素,返回 true
          return true;
        } catch (NoSuchElementException e) {
        //若找不到元素,返回 false
          return false;
        }
    }
```

也可以利用 sleep 封装一个计时器,完成等待操作,如代码清单 11-24 所示。

代码清单 11-24   通过 sleep 计时器实现隐式等待

```java
@Test(description = "sleep 封装")
public static void testwaitTimer( int units, int mills ) {
    DecimalFormat df = new DecimalFormat("###.##");
    double totalSeconds = ((double)units*mills)/1000;
    System.out.print("Explicit pause for " + df.format(totalSeconds) + " seconds divided by " + units + " units of time: ");
    try {
```

```java
        Thread.currentThread();
        int x = 0;
        while( x < units ) {
            Thread.sleep( mills );
            System.out.print("." );
            x = x + 1;
        }
        System.out.print('\n');
    } catch ( InterruptedException ex ) {
        ex.printStackTrace();
    }
}
```

隐式等待方式（implicitlyWait）是指在尝试发现某个元素的时候，如果没能立刻发现，就等待固定长度的时间。默认设置是 0s，如代码清单 11-25 所示。

代码清单 11-25　　　　　　　　　　implicitlyWait 实现隐式等待

```java
@Test(description = "测试显示等待")
public void testImplicitlyWait(){
    //识别"美食"图标
    MobileElement meishiElement = (MobileElement) driver.findElement(By.xpath("//android.support.v7.widget.RecyclerView/android.widget.RelativeLayout[1]/android.widget.ImageView"));
    //单击"美食"图标，跳转到"美食"界面
    meishiElement.click();
    //设置全局等待时间最大为30s
    driver.manage().timeouts().implicitlyWait(30, TimeUnit.SECONDS);
    //查找"深***"
    try{
        //查找"深***"
        driver.findElement(By.xpath("//android.widget.TextView[@text='深***']"));
    }catch(NoSuchElementException e){
        //如果控件没有找到，则测试用例执行失败
        Assert.fail("没有找到控件");
    }
}
```

11.16 显示等待方法

在自动化测试的过程中，很多窗体内的数据，需要等待一会儿，才能加载完数据，才能出现一些元素，Driver 才能操作这些元素。另外，做一些操作，本身可能也需要等待一会儿才有数据显示。

不管是否加载完成，隐式等待都会等待特定的时间，它会让一个正常响应的应用的测试变慢，增加了整个测试执行的时间。比如有些控件可能数据较多，需要较长时间才可以加载完成，但是其他控件加载很快，把它们都设置成固定等待时间，将会造成大量时间的浪费。因此，合理地设置时间等待是非常必要的。

Appium 中提供了 AppiumFluentWait 来实现显示等待。AppiumFluentWait 继承自 FluentWait。这个类能支持一直等待知道特定的条件出现，使用 AppiumFluentWait 可以设置最大等待时间、等待的频率等，如代码清单 11-26 所示。

代码清单 11-26   显示等待

```
@Test(description = "测试 FluentWait")
public void testFluent(){
//识别美团图标
MobileElement meituan = (MobileElement) driver.findElement(By.xpath("// android.support.v7.widget.RecyclerView/android.widget.RelativeLayout[1]/android.widget.ImageView"));
//创建 AppiumFluentWait 对象
new AppiumFluentWait<MobileElement>(meituan)
//最长等待时间为 10s
.withTimeout(10, TimeUnit.SECONDS)
//每隔 100ms 判断一次元素的文本值是否为"深***"
.pollingEvery(100,TimeUnit.MILLISECONDS)
.until(new Function<MobileElement,Boolean>(){
@Override
public Boolean apply(MobileElement element) {
return element.getText().endsWith("深***");
    }
});
}
```

AppiumFluentWait 的 until 的参数可以是 Predicate，也可以是 Function。这里使用的是 Function。因为 Function 的返回值种类较多，可以为 Object 或者 Boolean 类型，而 Predicate 只能返回 Boolean 类型。

11.17 在编程中处理 adb 命令

在对 App 进行性能测试时，如获取 CPU 信息的命令为 adb shell dumpsys cpuinfo packagename。在 selendroid-test-app-0.15.0.apk 实例中，要获取 CPU 的性能指标，编写的代码如代码清单 11-27 所示。

代码清单 11-27　　　　　　　　获取 CPU 的性能指标

```java
public static void GetCpu(String packageName) throws IOException {
    Runtime runtime = Runtime.getRuntime();
    Process proc = runtime.exec("adb shell dumpsys cpuinfo $"+packageName);
    try {
        if (proc.waitFor() != 0) {
            System.err.println("exit value = " + proc.exitValue());
        }
        BufferedReader in = new BufferedReader(new InputStreamReader(
            proc.getInputStream()));
        String line = null;
        String totalCpu = null;
        String userCpu = null;
        String kernalCpu = null;
        while ((line = in.readLine()) != null) {
            if(line.contains(packageName)){
                System.out.println(line);
                totalCpu = line.split("%")[0].trim();
                userCpu = line.substring(line.indexOf(":")+1,line.indexOf("% user")).trim();
                kernalCpu = line.substring(line.indexOf("+")+1, line.indexOf("% kernel")).trim();
                System.out.printf("totalCpu 的值为: %s%n", totalCpu);
                System.out.printf("userCpu 的值为: %s%n", userCpu);
                System.out.printf("kernalCpu 的值为: %s%n", kernalCpu);
            }
```

```
            }
        } catch (InterruptedException e) {
            System.err.println(e);
        }finally{
            try {
                proc.destroy();
            } catch (Exception e2) {

            }
        }
```

输出结果如图 11-7 所示。

```
<terminated> test_getcpu [Java Application] C:\Program Files\Java\jre1.8.0_144\bin\javaw.exe (2017-9-27 下午1:50:59)
 0.9% 20850/io.selendroid.testapp: 0.6% user + 0.2% kernel / faults: 148 minor
totalCpu的值为: 0.9
userCpu的值为: 0.6
kernalCpu的值为: 0.2
```

图 11-7 CPU 性能指标

在实际的测试过程中可以多次调用上述代码，以获取不同阶段的 CPU 值。其他性能指标的获取方法类似。

11.18　区分 WebElement、MobileElement、AndroidElement 和 iOSElement

在 Appium 自动化测试中，可能有些初学者会对获取控件元素对象的类型存在疑惑，不知道在什么情况下使用什么类型。下面将介绍控件元素对象类型的区别。

- WebElement 可以使用所有的 Selenium 命令。
- MobileElement 属于 Appium，继承自 WebElement，但是又增加了一些 Appium 特有的功能（如 Touch 手势）。
- AndroidElement 和 iOSElement 实现了 WebElement 接口方法，并增加了一些 Android 和 iOS 特有的功能（如 findByAndroidUiAutomation）。

根据待测手机操作系统平台，可以选择不同的应用，或者根据是否跨平台进行选择。

11.19　区分 RemoteWebDriver、AppiumDriver、AndroidDriver 和 iOSDriver

在 Appium 自动化测试中，可能有些初学者会对创建什么类型的驱动产生困惑，本节将介绍各个驱动类型的区别。

- RemoteWebDriver：这个驱动来自于 Selenium，可以使执行测试的机器和发送测试命令的机器独立开来，中间存在网络请求。Appium 是基于客户端/服务器的，所有 RemoteWebDriver 可以直接初始化会话。但是一般不建议使用，Appium 提供了其他驱动，可能在使用上更加方便。
- AppiumDriver：继承自 RemoteWebDriver，但是增加了一些特有的功能（如上下文切换）。
- AndroidDriver：继承自 AppiumDriver，但是增加了一些特有的功能，如 openNtificutions 方法，只有在 Android 设备或者 Android 模拟器上才使用这个驱动。
- iOSDriver：继承自 AppiumDriver，但是增加了一些特有的功能，只有在 iOS 设备或者 iOS 模拟器上才使用这个驱动。

在实际的使用场景中，根据手机操作系统不同，建议直接使用 AndroidDriver 或者 iOSDriver。

11.20　在代码中启动服务器

在 Appium 测试执行时，需要手动启动 Appium 服务器。在一些并行测试场景下，要启动多个 Appium 服务器，如果在代码中未使用 driver.quit 关闭服务器，或者存在其他一些异常，就会出现会话无法创建的情况。Appium 官网提供了 AppiumDriverLocalService 来完成 Appium 服务器的启动和关闭。这一节讲述如何设置 Appium 服务器的启动和关闭，可以根据项目要求进行集成。

使用 AppiumDriverLocalService 的前提条件有以下两个。

- 安装 Node.js 7 以上版本。
- 通过 npm 安装 Appium 服务器。

具体的操作如下。

（1）如果没有指定参数，实现方式如代码清单 11-28 所示。

代码清单 11-28　　　　　　　　未指定参数

```java
import io.appium.java_client.service.local.AppiumDriverLocalService;
...
  AppiumDriverLocalService service = AppiumDriverLocalService.buildDefaultService();
  service.start();
  ...
  service.stop();
```

本地环境中可能会在这一步报错。

```java
AppiumDriverLocalService service = AppiumDriverLocalService.buildDefaultService();
```

这个问题在 UNIX/Linux 下面比较常见，可能是因为使用的 node.js 实例与环境变量设置的实例不是同一个，也有可能是 Appium node 服务导致的（Appium.js 版本小于等于 1.4.16，Main.js 版本大于等于 1.5.0）。在这种情况下，建议用户设置 NODE_BINARY_PATH（Windows 操作系统下指 node.exe 所在路径，Linux/Mac OS 下指 node 所在路径）和 APPIUM_BINARY_PATH（Appium.js 和 Main.js 的执行路径）到环境变量中，也可以在程序中指定。

```java
//appium.node.js.exec.path
System.setProperty(AppiumServiceBuilder.NODE_PATH ,
"the path to the desired node.js executable");


System.setProperty(AppiumServiceBuilder.APPIUM_PATH ,
"the path to the desired appium.js or main.js");
AppiumDriverLocalService service = AppiumDriverLocalService.buildDefaultService();
```

（2）指定参数，如代码清单 11-29 所示。

代码清单 11-29　　　　　　　　指定参数

```java
import io.appium.java_client.service.local.AppiumDriverLocalService;
import io.appium.java_client.service.local.AppiumServiceBuilder;
import io.appium.java_client.service.local.flags.GeneralServerFlag;
...
```

```
AppiumDriverLocalService service = AppiumDriverLocalService.
buildService(new AppiumServiceBuilder().
withArgument(GeneralServerFlag.TEMP_DIRECTORY,
            "The_path_to_the_temporary_directory"));
```

或者

```
import io.appium.java_client.service.local.AppiumDriverLocalService;
import io.appium.java_client.service.local.AppiumServiceBuilder;
import io.appium.java_client.service.local.flags.GeneralServerFlag;
...

AppiumDriverLocalService service = new AppiumServiceBuilder().
withArgument(GeneralServerFlag.TEMP_DIRECTORY,
            "The_path_to_the_temporary_directory").build();
```

需要导入以下 3 个包。

```
io.appium.java_client.service.local.flags.GeneralServerFlag
io.appium.java_client.service.local.flags.AndroidServerFlag
io.appium.java_client.service.local.flags.iOSServerFlag
```

（3）定义参数。

在有些情况下可能需要使用一些特殊的端口（指定端口）。

```
new AppiumServiceBuilder().usingPort(4000);
```

或者使用那些未使用的端口。

```
new AppiumServiceBuilder().usingAnyFreePort();
```

使用其他的 IP 地址。

```
new AppiumServiceBuilder().withIPAddress("127.0.0.1");
```

确定日志文件。

```
import java.io.File;
  ...


new AppiumServiceBuilder().withLogFile(logFile);
```

Node.js 执行路径，如代码清单 11-30 所示。

代码清单 11-30　　　　　　　　Node.js 执行路径

```
import java.io.File;

...

new AppiumServiceBuilder().usingDriverExecutable(nodeJSExecutable);
```

Main.js 执行路径，如代码清单 11-31 所示。

代码清单 11-31　　　　　　　　Main.js 执行路径

```
import java.io.File;

...

//appium.js is the full or relative path to
//the appium.js (v<=1.4.16) or maim.js (v>=1.5.0)
new AppiumServiceBuilder().withAppiumJS(new File(appiumJS));
```

确定服务器端的 Desired Capabilities，如代码清单 11-32 所示。

代码清单 11-32　　　　　确定服务器端的 Desired Capabilities

```
DesiredCapabilities serverCapabilities = new DesiredCapabilities();
...//the capability filling

AppiumServiceBuilder builder = new AppiumServiceBuilder().
withCapabilities(serverCapabilities);
AppiumDriverLocalService service = builder.build();
service.start();
...
service.stop();
```

11.21　PageFactory 注解

第 8 章中使用了 Page Object 和 PageFactory 两种设计模式。这一节将详细阐述 Appium 官方关于 Page Object 和 PageFactory 的使用，并通过实例加深对它们的认识，以便在实际使用中对这些概念不会产生疑惑并能灵活地根据需求进行设置。更复杂的使用场景参考官方文档。

（1）如代码清单 11-33 所示，默认设置为 WebElement 或 WebElement 数组，注释方式使用 FindBy，元素类型为 WebElement。

代码清单 11-33　　　　　　　　　　FindBy 实例

```
import org.openqa.selenium.WebElement;
import org.openqa.selenium.support.FindBy;
...

@FindBy(someStrategy) //for browser or web view html UI
//also for mobile native applications when other locator strategies are not defined
WebElement someElement;

@FindBy(someStrategy) //for browser or web view html UI
//also for mobile native applications when other locator strategies are not defined
List<WebElement> someElements;
```

（2）指定具体的定位策略。如代码清单 11-34 所示，根据 Desired Capability 中设置的 automationName 自动化测试引擎的值，针对移动原生应用（Native App），分别使用"@AndroidFindBy""@SelendroidFindBy"和"@iOSFindBy"进行注解，元素类型为 AndroidElement、RemoteWebElement 以及 IOSElement。

代码清单 11-34　　　　　　　　　　指定具体的定位策略

```
import io.appium.java_client.android.AndroidElement;
import org.openqa.selenium.remote.RemoteWebElement;
import io.appium.java_client.pagefactory.*;
import io.appium.java_client.ios.IOSElement;

@AndroidFindBy(someStrategy) //for Android UI when Android UI automator is used
AndroidElement someElement;

@AndroidFindBy(someStrategy) //for Android UI when Android UI automator is used
List<AndroidElement> someElements;

@SelendroidFindBy(someStrategy) //for Android UI when Selendroid automation is used
RemoteWebElement someElement;

@SelendroidFindBy(someStrategy) //for Android UI when Selendroid automation is used
```

```
List<RemoteWebElement> someElements;

@iOSFindBy(someStrategy) //for iOS native UI
IOSElement someElement;

@iOSFindBy(someStrategy) //for iOS native UI
List<IOSElement> someElements;
```

(3)跨平台的原生 App 测试实例,如代码清单 11-35 所示。针对原生 App,使用"@AndroidFindBy"和"@iOSFindBy"同时进行注解。元素的类型为 MobileElement。

代码清单 11-35　　　　　　　跨平台的原生 App 测试实例

```
import io.appium.java_client.MobileElement;
import io.appium.java_client.pagefactory.*;

@AndroidFindBy(someStrategy)
@iOSFindBy(someStrategy)
MobileElement someElement;

@AndroidFindBy(someStrategy) //for the crossplatform mobile native
@iOSFindBy(someStrategy)    //testing
List<MobileElement> someElements;
```

(4)全平台的测试实例,如代码清单 11-36 所示。其中使用"@FindBy""@AndroidFindBy"以及"@iOSFindBy"同时进行注解。元素的类型为 RemoteWebElement。

代码清单 11-36　　　　　　　全平台的测试实例

```
import org.openqa.selenium.remote.RemoteWebElement;
import io.appium.java_client.pagefactory.*;
import org.openqa.selenium.support.FindBy;

//the fully cross platform examle
@FindBy(someStrategy) //for browser or web view html UI
@AndroidFindBy(someStrategy) //for Android native UI
@iOSFindBy(someStrategy)  //for iOS native UI
RemoteWebElement someElement;

//the fully cross platform examle
@FindBy(someStrategy)
```

```
@AndroidFindBy(someStrategy) //for Android native UI
@iOSFindBy(someStrategy)    //for iOS native UI
List<RemoteWebElement> someElements;
```

（5）用 Chained 或者 Possible 定位方式。

- Chained 定位方式。使用 "@FindBys" "@AndroidFindBys" 和 "@iOSFindBy" 进行注解。元素内容通过多种定位方法找到。FindBys 相当于在多种定位方式中取交集，如 "@FindBys({@FindBy(someStrategy1)" "@FindBy(someStrategy2)})" 相当于首先根据 someStrategy1 找到对应元素，然后在这些元素中通过 someStrategy2 再次查找元素，这类似于 driver.findelement(someStrategy1). findelement(someStrategy2)，如代码清单 11-37 和代码清单 11-38 所示。

代码清单 11-37　　　　　　　　Chained 定位方式之一

```
import org.openqa.selenium.remote.RemoteWebElement;
import io.appium.java_client.pagefactory.*;
import org.openqa.selenium.support.FindBys;
import org.openqa.selenium.support.FindBy;

@FindBys({@FindBy(someStrategy1), @FindBy(someStrategy2)})
@AndroidFindBy(someStrategy1) @AndroidFindBy(someStrategy2)
@iOSFindBy(someStrategy1) @iOSFindBy(someStrategy2)
RemoteWebElement someElement;

@FindBys({@FindBy(someStrategy1), @FindBy(someStrategy2)})
@AndroidFindBy(someStrategy1) @AndroidFindBy(someStrategy2)
@iOSFindBy(someStrategy1) @iOSFindBy(someStrategy2)
List<RemoteWebElement> someElements;
```

代码清单 11-38　　　　　　　　Chained 定位方式之二

```
import org.openqa.selenium.remote.RemoteWebElement;
import io.appium.java_client.pagefactory.*;
import org.openqa.selenium.support.FindBys;
import org.openqa.selenium.support.FindBy;

import static io.appium.java_client.pagefactory.LocatorGroupStrategy.CHAIN;

@HowToUseLocators(androidAutomation = CHAIN, iOSAutomation = CHAIN)
```

```
@FindBys({@FindBy(someStrategy1), @FindBy(someStrategy2)})
@AndroidFindBy(someStrategy1) @AndroidFindBy(someStrategy2)
@iOSFindBy(someStrategy1) @iOSFindBy(someStrategy2)
RemoteWebElement someElement;

@HowToUseLocators(androidAutomation = CHAIN, iOSAutomation = CHAIN)
@FindBys({@FindBy(someStrategy1), @FindBy(someStrategy2)})
@AndroidFindBy(someStrategy1) @AndroidFindBy(someStrategy2)
@iOSFindBy(someStrategy1) @iOSFindBy(someStrategy2)
List<RemoteWebElement> someElements;
```

- Possible 定位方式。如代码清单 11-39 所示，这种定位方式指使用"@FindAll" "@AndroidFindAll"和"@iOSFindAll"进行注解。FindAll 相当于在多种定位方式中取并集，如"@FindAll{@FindBy(someStrategy1)"，"@FindBy(someStrategy2)})"相当于取到所有符合 someStrategy1 和 someStrategy2 的元素。

代码清单 11-39　　　　　　　　　Possible 定位方式

```
import org.openqa.selenium.remote.RemoteWebElement;
import io.appium.java_client.pagefactory.*;
import org.openqa.selenium.support.FindBy;
import org.openqa.selenium.support.FindByAll;

import static io.appium.java_client.pagefactory.LocatorGroupStrategy.ALL_POSSIBLE;

@HowToUseLocators(androidAutomation = ALL_POSSIBLE, iOSAutomation = ALL_POSSIBLE)
@FindAll{@FindBy(someStrategy1), @FindBy(someStrategy2)})
@AndroidFindBy(someStrategy1) @AndroidFindBy(someStrategy2)
@iOSFindBy(someStrategy1) @iOSFindBy(someStrategy2)
RemoteWebElement someElement;

@HowToUseLocators(androidAutomation = ALL_POSSIBLE, iOSAutomation = ALL_POSSIBLE)
@FindAll({@FindBy(someStrategy1), @FindBy(someStrategy2)})
@AndroidFindBy(someStrategy1) @AndroidFindBy(someStrategy2)
@iOSFindBy(someStrategy1) @iOSFindBy(someStrategy2)
List<RemoteWebElement> someElements;
```